SpringerBriefs in Compu

CW01507138

For further volumes:
http://www.springer.com/series/10028

Alan Moran

Agile Risk Management

 Springer

Alan Moran
Zurich
Switzerland

ISSN 2191-5768   ISSN 2191-5776 (electronic)
ISBN 978-3-319-05007-2   ISBN 978-3-319-05008-9 (eBook)
DOI 10.1007/978-3-319-05008-9
Springer Cham Heidelberg New York Dordrecht London

Library of Congress Control Number: 2014931762

Springer is part of Springer Science+Business Media (www.springer.com)

Foreword

Agile is no longer a hype. It has matured—we even have an Agile Maturity Model! It is being used across the globe and continues to be innovated, overcoming many old myths as well as some new ones. However, our work is not complete and it will always demand of us that we keep up with new developments in technology, society and our working environment. One of these areas is agile risk management.

The most used method of risk management I know is people just writing risks down! With pride people will show you their most brilliant Excel sheets with fifteen to twenty columns that present information on what they perceived as risks in a given situation. Rarely do I meet people who actually *manage* risk, let alone manage it in an *agile* way!

This brings us to an important question, what is agile risk management? I have been discussing this with Alan and we agree on the fact that agile risk management means managing risks in such a way that it facilitates the agile interaction concept just like any other agile practice or technique.

Alan takes agile risk management a step further. He changes the perception of what a risk is and what it can bring to a project or a team. We have been over this ground again and again and each time more value has been added. He busts a few myths and brings new techniques to the table. I hope that you will not only enjoy this book but that it will also help you in your work and give you new insights. I congratulate Alan on this work and hope to see more like it soon!

Arie van Bennekum

Preface

This book is a critical analysis of the practice of risk management in agile software development projects. Risk, defined in terms of uncertainty relating to project objectives, is treated both as a threat and as an opportunity wherein the pitfalls and rewards that underpin project success lie. Although the agile community frequently cites risk management, research suggests that risk is often narrowly framed and at best implicitly treated, which in turn leads to an inability to make informed decisions concerning risk and reward and a poor understanding about when to engage in risk-related activities. Moreover, the absence of reference to enterprise risk management means that project managers are unable to clearly articulate, scope or tailor their projects in line with the wider expectations of the organisation.

Yet the agile approach, with its rich toolset of techniques, is more than equipped to effectively and efficiently deal with the risks that arise in projects. In this book we endeavour to address the above issues by proposing an agile risk management process derived from classical risk management but adapted to the circumstances of agile projects. We thus express the agile approach to risk management and illustrate its application to selected methodologies (XP, Scrum and DSDM) chosen on account of their varying foci on the software development process and their attitudes towards risk. Though our interest lies in the software development process, much of what we say could be applied to other types of IT projects.

Audience

This book is intended for those directly involved in agile software development who share a concern for how risk should be managed. The primary interest groups include project and risk managers, agile practitioners and general IT managers. Whilst we do not presume a thorough background in risk management, we do assume some basic level of familiarity with or exposure to agility. Where appropriate we refer the reader to more detailed sources in the literature.

Overview

We begin in the chapter "Agile Software Development" with an initial survey of agility focusing on those aspects that are of relevance later in the book and use this opportunity to introduce our three main methodologies (i.e., XP, Scrum and DSDM). We characterize the cyclical nature of iterative development and incremental delivery in terms of *agile charting* (and related notions such as *slicing*, *clock facing* and *escape velocity*) and show how this tool can be used to facilitate communication and improve understanding within an agile team. We conclude with some remarks concerning the current state of agility and comment briefly on the management perspective.

In the "Project Risk Management" chapter we formally define project risk and conduct a comprehensive survey of project risk management as it is understood by risk managers. We illustrate the consensus view by synthesizing best practices into a generic model of project risk management before moving on to the notion of enterprise risk management. In effect this sets the standard and defines the core concepts that agile risk management must embrace if we are to seriously apply risk management in agile projects. The reader already familiar with the details of project risk management may choose to skim over this chapter.

In the chapter "Agile Risk Management", we first explore how risk is perceived and identify some of the shortcomings of agile methodologies before proposing an agile risk management process that is loosely based on traditional project risk management, though we introduce a number of adaptations that make it more meaningful in the context of agile projects. This process is concerned with how to risk scope a project and how to interpret this in the context of the wider risk environment by introducing the notion of a *risk driver map*. We then use agile charting to explore how a methodology can be risk tailored at the project level. In our treatment of risk management we explain a couple of techniques that can be used to identify risks during iteration planning and then go on to explain the options available to risk managers and the principles that underpin them. We introduce a number of tools such as a *risk list* and show how risks can be treated with a combination of *risk tasking*, *risk techniqueing* or *contingency planning*. We show how to make risks visible using a *risk modified Kanban board* and move on to describing a risk reporting technique using *risk burndowns*. We acknowledge the systemic nature of risk, *iteration residual risk*, and how to measure the effectiveness of risk management in terms of the *iteration residual risk ratio*.

In "Applying Agile Risk Management" we illustrate the application of the agile risk management process to our chosen methodologies. We critically review each methodology and describe its chief characteristics and level of maturity in relation to risk. From there we offer concrete advice and guidance on how to conduct risk management and relate our suggestions to existing artefacts and practices found within the respective methodologies.

Our final chapter on "Enterprise Agility" notes the rise of frameworks (including DAD and SAFe) that attempt to scale agile practices to the enterprise and we evaluate their contribution to agile risk management. We note an absence of reference to enterprise risk management though there are indications of a growing awareness and maturity.

Terminology

Throughout we strive towards simplicity, clarity and neutrality in our use of terminology and for reasons of personal taste we often prefer the term "agility" over "agile" (e.g., "enterprise agility" rather than "enterprise agile"). We seek to use neutral language that is already widely accepted or understood within the agile community. Thus we refer to "daily stand-ups" (rather than the more methodologically specific "Daily Scrum"), "Kanban (board)" (rather than "Scrum-ban") and "backlog" (rather than the "product/Sprint backlog" of Scrum or the "prioritized requirements list" of DSDM). We trust that the context will render clear what is meant by our use of the terms and that no bias towards a specific methodology be inferred through our choice of nomenclature. Needless to say some terms are simply applied differently according to methodology so that although we use "iteration" in the manner already defined earlier, we appreciate that this term is used in a broader sense in Scrum and a narrower sense in DSDM. Instead we respect our mutual differences and endeavour to make our language more precise where appropriate.

Acknowledgments

Though this book was born of efforts by the author to integrate risk management practices over a period of many years of setting up and working with agile processes, a truly deep understanding of how agility really works can only be achieved by working together with and learning from others. We would like to extend our thanks to all who directly or indirectly contributed to this book through their discussions, feedback and comments and through the exchange of experiences based on mutual respect and tolerance. Special thanks is afforded to our reviewers Scott Amber, Arie van Bennekum, Jutta Eckstein, Julia Godwin, Margaret Stewart and Patrick Verheij whose insightful remarks and comments helped validate and clarify the ideas raised in this book. Finally, since nothing would have been possible without the love and support of Helen, Markus and Patrick it is to them that I owe an unrepayable debt!

Contents

Agile Software Development

Abstract Looking back at the origins of agile software development we provide an overview of its unique blend of supple practices and vivid cultural facets that contrast so sharply with the plan-driven world of Waterfall methodologies. We discuss the iterative and incremental nature of agility and introduce a tool, agile charting, that can be used to facilitate communication within agile teams. Against this backdrop we introduce and compare the three methodologies (i.e., XP, Scrum and DSDM) used throughout this book to illustrate our application of agile risk management. We conclude with a glimpse at the state of affairs of agility today and at the management perspective on agile project management thereby setting the tone for the remainder of the book.

Agile Defined

Agility, as a concept, came to prominence in the late 1990s through the endeavours to address perceived difficulties with existing software development processes that were rooted in, and owed their rigidity to, plan-driven practices. Advocates of agility promoted the notion that project uncertainties should be embraced and sought to balance planning and control with execution and feedback. Accordingly, agile projects exhibit features of open communication amongst heterogeneous stakeholders, emergent behaviour within self-organizing teams and a culture of openness and learning (Cockburn 2007). Central to agile software development are the notions of iterative development and incremental delivery based on shared values enshrined in the agile manifesto reproduced here for convenience.

A. Moran, *Agile Risk Management*, SpringerBriefs in Computer Science, DOI: 10.1007/978-3-319-05008-9_1, © The Author(s) 2014

We are uncovering better ways of developing software by doing it and helping others do it.
Through this work we have come to value:

Individuals and interactions over processes and tools
Working software over comprehensive documentation
Customer collaboration over contract negotiation
Responding to change over following a plan

That is, while there is value in the items on the right, we value the items on the left more.
(Beek et al. 2001a).

This manifesto expresses the beliefs that interactions amongst project team members and their customers should underpin efforts to create working software in a flexible manner. Equally important are the twelve principles (Beck et al. 2001b) that emphasize continuous delivery, an attitude of embracing change, frequent delivery of functional components, daily interaction with business stakeholders, empowerment through trust and support, direct communication, measurement of progress through functional software, sustainability, continuous excellence of design, simplicity through minimalism, self-organization and team reflection. Agility is as much a cultural stance on the process of software development, as it is a set of practices and values. The discipline and focus needed to practice agility is frequently underestimated (Boehm and Turner 2009) and has even been famously described as "hard and disruptive" (Schwaber 2006) by one of its leading proponents.

Agility embodies a rich texture of humanist (e.g., cognitive, social and interpersonal), organizational (e.g., managerial and cultural) and technological (e.g., practical and technical aspects) traditions (Hazzan and Dublinsky 2008) and there already exists a mature body of literature concerning both its culture (Cockburn 2007) and its practices (Coplien and Harrison 2005; Cohn 2010; Derby and Larsen 2009). There are numerous comparative surveys of agile methodologies (Larman 2003; Boehm and Turner 2009; Cockburn 2007) that highlight both the commonality rooted in the manifesto (and its principles) together with the uniqueness of focus and purpose with which each approach is endowed. It is not our purpose to explore the fine detail of all these methodologies, but rather we limit ourselves to a detailed discussion of three popular approaches (XP, Scrum and DSDM[1]) and to the techniques they employ. For convenience, the techniques are described in Appendix A and we encourage the reader to refer to this list when we mention a specific technique (e.g., Kanban[2] boards, burndown charts). Throughout this book we use the term *methodology* to refer a "system of methods and principles used in a particular discipline" (McKeown and Holmes 2009) consistent with the notion that it embodies not only practices but also principles, roles, artefacts and phases. We refer the interested reader to (Cockburn 2007) for a detailed discussion of this topic.

---

[1] Throughout this work when mentioning DSDM we will invariably be also referring to its project management cousin, AgilePM®, developed by the DSDM Consortium and accredited by the APMG International™.

[2] Throughout this book when we mention "Kanban" we usually mean just the board based technique. Kanban in the wider sense is, however, a methodology in its own right.

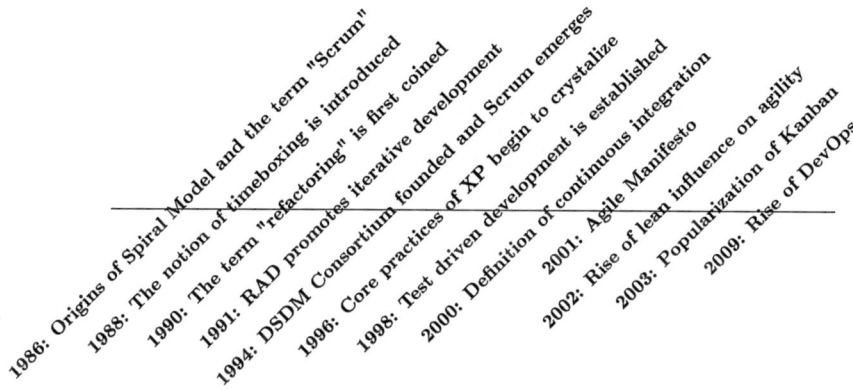

Fig. 1 Agile Timeline. Adapted from (Agile Alliance 2013). Published with kind permission of © Alan Moran 2013. All Rights Reserved

The origins of agility are rooted in part in the Japanese manufacturing and industrial sectors to which many an agile concept owes its heritage. These include the visual control concept found in the Toyota Production System that later anticipated agile information radiators, the Kanban charts used in agile task assignment and tracking, and the continual influence of lean thinking on agility today (Poppendieck and Poppendieck 2003). The synthesis of Eastern and Western thinking so persuasively laid out in (Takeuchi and Nonaka 2013), from the authors that introduced the term "Scrum", reflects the spirit in which the agile manifesto itself was conceived. The timeline depicted in Fig. 1 marks out some of the key elements of relevance to this book[3] and is adapted and extended from (Agile Alliance 2013).

Iterations and Increments

From its earliest inception, agility was understood as more of an evolution rather than a revolution. Its present day form is heavily influenced by the school of iterative development and incremental delivery that was by the 1980s already well established. Indeed movements such as Rapid Application Development and the IBM Rational Unified Process gave birth to their modern day agile counterparts in the form of Dynamic Systems Development Method and Agile Unified Process respectively. Embedded in all of these is the notion of the Software Development Lifecycle[4] (SDLC) which refers to a generic model of software development comprising of phases broadly divided into the soliciting and management of requirements, technical analysis and design, implementation of software solutions, validation and verification

---

[3] Our timeline omits many significant events to which we refer the interested reader to (Agile Alliance 2013).

[4] The term *Systems* Development Lifecycle is also in widespread use.

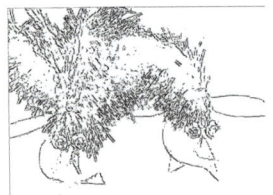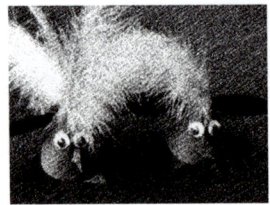

Fig. 2 Metaphor for the Iterative Evolution of a Solution. Published with kind permission of © Alan Moran 2013. All Rights Reserved

of software artefacts and concludes with the deployment and maintenance of the software solution. Implementations of the SDLC vary from the Waterfall model (Benington 1983; Royce 1970), that advocates transition through the model in terms of a strictly linear process with each phase being commenced only when the previous phase is complete, to the agile approach that promotes rapid and repeated transitions through all phases (some of which could even be said to be merged) resulting in the iterative derivation of a solution. It is interesting to note that the Waterfall method, which was adapted from the manufacturing industry, brought with it the assumption that changes late in the development process were prohibitively costly. This has been rigorously challenged by agilists who argue that postponement of design decisions until sufficient information is available is possible and endorse practices such as refactoring[5] to cope with change. Between these the two viewpoints of Waterfall and agility, which were conceived approximately three decades apart, can be found a plethora of variants comprising of both plan-driven and humanistic influences. Indeed agility finds itself today in a continual state of evolution with latter day influences emerging once again from industry in the form of lean development practices as described in (Poppendieck and Poppendieck 2003). The iterative and incremental practices, atop of which agility is built, laid the foundations for feedback mechanisms used to cope with change that have become the hallmark of agility.

For the purposes of this book, *iterative development* refers to the traversal of the entire software development lifecycle with the aim of producing a self-contained, tested and partially functional product within a fixed timeframe. During successive iterations the product is further refined thereby enabling lessons learned from earlier iterations to be fed back into the process. *Incremental delivery* refers to the packaging and deployment of a product artefact that can be meaningfully used by the customer. Typically an increment will require several iterations to complete and several increments are necessary before the final product is delivered. Borrowing from a metaphor appearing in (Patton 2008) Fig. 2 describes the iterative manner in which an artefact evolves. Though the general idea is clear from the outset, details emerge gradually over time allowing participants to incorporate feedback back into the production process. At each stage there is a gaining of understanding and an adding of detail which can be clearly demonstrated and delivered to the customer.

---

[5] We remind the reader to refer to Appendix A for a definition of this and other agile techniques.

Some methodologies, notably Scrum, merge these two concepts into one referring to the process of "iterative and incremental development" (Schwaber 2004). Thus consistent with our view of iterative development, Jeff Sutherland, a leading advocate of Scrum, describes iteration as the act of traversal of the whole during each pass of which the product gradually comes into focus. For him increments are concerned with the notion that "incremental development is iterating on the whole thing" and that each iteration should conclude with a "minimal useable feature set that is potentially shippable" indicating that this implies that code must satisfy the following definition of done for the increment

> thoroughly tested, well-structured and well-written code that has been built into an executable and that the user operation of the functionality is documented, either in Help files or in user documentation (Schwaber 2004).

though he later concedes that he "could have been clearer on what 'potential ship-pable software' means" (Sutherland 2010). There is thus a suggestion that incre-mental activity embodies the characteristics of a deployable artefact though Scrum leans towards the use of this term in the substantive rather than adjectival form e.g., each iteration "delivers a fully functional increment" (Sutherland 2010). In essence our use of the term, increment, reflects a conceptual designation that implies the potential release of a deployment artefact. In reality there is little consensus in the agile community concerning the precise definition of an increment or indeed where the boundaries of agility lie. For example, owing to the structure of most organiza-tions, it ought not be assumed that a product development team has full control over the means of deployment or of the dissemination of documentation (e.g., corporate design, accessibility). Moreover, it is ordinarily the case that some deployment activ-ities (e.g., training of users and service desk staff) are assumed by people outside of the project team. Whilst product focused methodologies (e.g., XP and Scrum) rarely consider what happens to the deliverable, other methodologies (e.g., DSDM, DAD, SAFe) consider it within their remit to manage the entire delivery process from inception to transition. Accordingly, we allow ourselves to make this important con-ceptual distinction between these two related but different activities but accept that the terms iteration and increment will have their own connotations within specific methodologies.

Agility in Practice

Agile teams tend to be small[6] comprising of heterogeneous "generalizing specialists" (Ambler 2003) capable of engaging in several distinct types of work (e.g., analysis, development, testing). Customer representatives are expected to be highly engaged, attend planning and demonstration events and be available on short notice should the solution team require their input. In this context the acronym *CRACK*,[7] first coined

---

[6] Team sizes are consistently recommended to be in the range 7 ± 2 (Miller 1956).

[7] Collaborative, Representative, Authorized, Committed and Knowledgeable.

in (Boehm and Turner 2009), sets the tone for what is expected of such representatives. This is in stark contrast to pre-agile approaches where business and development teams tended to be separated with relatively little contact beyond exchange of requirements and specifications. Agile methodologies employ a wide range of techniques and there is considerable overlap amongst the common methodologies both in their interpretation and their application of them. Some of these practices work well in concert (e.g., refactoring and continuous integration) whilst others might be considered complementary approaches to tackling a problem (e.g., modelling and prototyping). There does appear, however, to be a core set of practices that are used by most agile teams (which we see later comprises of daily stand-ups, iteration and release planning, unit testing, retrospectives, continuous integration, automated builds and burndown charting) and enjoy a common interpretation, though the precise wording may vary by methodology. Thus whilst it is true that all methodologies bear their own interpretation of the agile manifesto, it is equally fair to say that they borrow heavily from each other.

Generally speaking a team must balance the need for adaptation (e.g., innovation) against the necessity of optimization. This suggests that lightweight methodologies service high adaptation and low optimization environments better, whereas heavier methodologies are to be found in low adaptation and high optimization contexts (Cockburn 2007). It ought to be noted, however, that an enterprise typically requires both. Indeed it has been argued that the stability provided by a mature process oriented organization creates the necessary conditions for an innovative agile environment to flourish (Stephen et al. 2011)—a view that is perhaps at odds with the accepted wisdom within agile communities.

Agile practices are best understood at the daily, iterative and incremental levels for which we will later introduce the agile charting tool (the curious can take a glance at Fig. 4 for details). A typical agile day begins with a stand-up meeting of the project team members. During the day code may be integrated into a shared repository perhaps using continuous integration practices thereby ensuring a tight feedback loop (e.g., unit testing may be performed as part of the integration thus providing an early warning system for developers). At the end of each day a complete build and deployment may be performed to assess the stability and readiness of the code as well as demonstrating working software. Technical practices performed on a daily basis tend to be highly automated (e.g., continuous integration).

Each iteration begins with a planning session (which might be preceded by a "grooming" which is essentially a feature triage ahead of the main planning event) during which estimates and priorities are set (see Cohn 2010 for a detailed discussion of this topic). Planning Poker, a simplified form of the Wideband Delphi method, is popularly used to gain consensus on estimates for project tasks. Throughout the iteration progress may be tracked in the form of burndown charts reflecting the high degree of transparency often found in agile working environments. In addition a Kanban board may be used to monitor the status of individual tasks. Towards the end of the iteration the project team invites the customer to test the code and to attend a demonstration of the work completed during the iteration. Concluding an iteration, the team reflects on its experiences (often referred to as a retrospective) and considers what might be done to improve the process (an excellent guide to such practices can be found in Derby and Larsen 2009). Iteration length typically varies between two to four weeks but remains fixed.

An increment requires high level planning often resulting in a feature list (the term "backlog" was popularized by XP and Scrum) that describes the requirements in the language of the customer at a level of detail commensurate with the information available (later subsets of this list becomes refined at the iteration level). Shared repository code is often isolated from the mainstream product code (a practice known as branching, itself a sensible risk management practice) though it is strongly recommended to instil a culture of frequent merging in order to avoid integration issues later. Acceptance testing of the evolving product usually occurs at all levels though its definitive character at the increment level is expected to complement the integration testing that occurs at the iteration and the unit testing in the daily cycle. Final delivery of a part (or complete) solution provides the welcome opportunity to celebrate fulfilment of customer requirements!

Comparing Methodologies

Today there are several well established agile methodologies in use, ranging from lightweight approaches that have a strong product development focus such as eXtreme Programming (Beck and Andres 2004) and Scrum, (Pichler 2010; Schwaber 2004), to heavier methodologies such as the project management focused Dynamic Software Development Method (DSDM 2010) or the model centric Agile Unified Process (Ambler 2012). Many other, lesser known, methodologies are also still moderately active including Feature Driven Development (Palmer and Felsing 2002), the Crystal family (Cockburn 2007) and Evo (Gilb 2005) to name but a few. Each methodology has its own culture, practices and language.[8] For example, both XP and Scrum express their principles in humanist terms (e.g., respect, courage) whereas DSDM may appeal to those more comfortable in a mature environments with a stronger structural focus as evidenced in its formulation of values (e.g., focus on business need, deliver on time, collaborate, demonstrate control). In this book we limit our focus to XP, Scrum and DSDM and use these to illustrate the application of agile risk management. Our choice is motivated by the belief that these three methodologies lie on a continuum in terms of ceremony and formality ranging from the minimalist technique-driven and engineering focused XP, through to the pragmatic product-project paradigm that is Scrum and on to comparatively formal project-centric DSDM. Each enjoys the following of a vibrant community and all are well documented (Beck and Andres 2004; Pichler 2010; Schwaber 2004; DSDM 2010). We now briefly comment on each of these methodologies though

---

[8] In the case of Evo a formal language, *planguage*, is even used to describe requirements and functional specifications!

we are circumspect in our discussions as we shall return to each in the "Applying Agile Risk Management" chapter.

Extreme Programming (XP) was refined out of the Chrysler Comprehensive Compensation System payroll system development methodology. These experiences found expression in 1999 (Beck and Andres 2004) and drew heavily on existing practices taken to the extreme. These include the writing of unit tests *prior* to implementing the solution as a means of clarifying what constitutes done and ensuring immediate feedback on progress towards and validation of project tasks (Beck 2002). The accusation levied against XP of being merely "old wine in new bottles" (Hikaka et al. 2005) seems unduly harsh since it fails to take into account the synthesis of practices and the cultural shift in software development that accompanied them. XP is therefore not without its critics (Boehm and Turner 2009; McBreen 2002) though criticisms are often qualified on grounds that misunderstanding can lead to misapplication. XP became hugely influential in the 1990s and its practices today constitute the essential tool set of every agile IT project.

Scrum can be traced back to its industrial heritage in 1986 (Takeuchi and Nonaka 1986) where it was conceived as a manufacturing development methodology that brought about "innovation continuously, incrementally and spirally" (Takeuchi and Nonaka 2013). These ideas regarded ambiguity during the development process as a strength, provoking new perspectives and challenging established ideas, rather than a weakness that must be managed through precise planning. Through the sharing of ideas and the embracing of heterogeneity of product teams, personal knowledge is encouraged to become organizational knowledge. Indeed, we begin to see here the hints of the genesis of agility and the ideas underpinning its manifesto. Drawing on sporting metaphors the "relay-race" of the Waterfall methodology was contrasted with the "scrum" of holistic agility (Takeuchi and Nonaka 1986), an expression later popularized in (Schwaber and Beedle 2001) which built on experiences gathered during the 1990s. It is interesting at this juncture to observe the merger of lean thinking, also taken from Japanese industry, resulting in the use of Kanban boards (co-opted in Scrum under the notion "Scrum-ban" though this is considered an optional element). Today Scrum enjoys widespread popularity and its maturity is reflected in the size of its community and the automated tool support of its practices (Altlassian 2013). During the 1990s Scrum became increasingly defined around a core model and set of practices, encouraged by some cross fertilization with XP.

The 1990s also saw the rise of Rapid Application Development (RAD) as a means to integrate planning into software development through the extensive use of prototyping and modelling (Martin 1991). It employed iterative practices that drew upon the Spiral Model (Boehm 1986) and other ideas which had been practised by IBM in the 1980s. The takeover in 2003 of the Rational Software Corporation, however, saw IBM's focus shift to the Unified Process and its latter day agile variant. This notwithstanding, in 1994 Dynamic Systems Development Method (DSDM), briefly marketed for a period under the name Atern®, emerged as an independent framework that has evolved over time to encompass a wider scope than one would traditionally associate with agile projects (e.g., inclusion of explicit quality, risk and configuration management). That DSDM influence is so visible in AgilePM®, suggests that it has

Table 1 Dimensions underpinning the comparision of agile methodologies

Dimension	Description
Principles	The core values that characterize the methodology and imbue it with meaning in the eyes of its practitioners
Roles	Distinct roles cited by the methodology. Note that several roles may be assigned to an individual and thus no conclusions should therefore be drawn regarding team size
Artefacts	The intermediate products generated and consumed by the process (omitting final project deliverables). The necessity to create indirect artefacts is a reliable indicator of the weight of a methodology
Practices	The techniques explicitly cited by the methodology as being core to the effective and efficient operation of the process
Phases	The distinct phases of the model underpinning the methodology through which the process must traverse

Fig. 3 Dimensional Comparison of XP, Scrum and DSDM. Published with kind permission of © Alan Moran 2013. All Rights Reserved

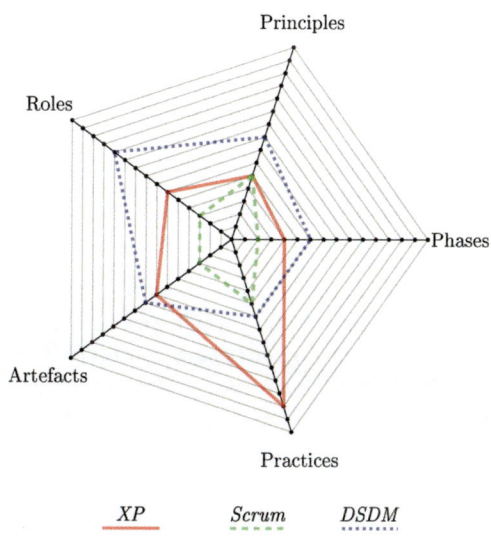

already established itself as the leading agile project management framework. Indeed, DSDM has made the case for embedding other product focused methodologies, in particular Scrum, within its framework (DSDM Consortium 2012b). Inspired by the desire to develop a methodology compatible with existing frameworks including ITIL®and PRINCE2®(DSDM Consortium 2010a), DSDM can be said to be rather heavy in terms of ceremony and formality when compared to other agile methodologies though this could equally be attributable to its richness in advice and guidance. We conclude our discussion of these methodologies with a comparison based on the dimensions listed in Table 1 and illustrate their differences in Fig. 3.

Needless to say each methodology elaborates on its practices to differing degrees of detail and thus a certain amount of interpretation is called for in order to make reasonable comparisons. In fact, some caution is called for when making direct comparisons based on the primary sources of these methodologies. For example Scrum describes the team (excluding the Scrum Master and Product Owner) as a role although in organizational terms this might be better described as a unit or function. Accordingly our analysis dips deeper using multiple sources to derive comparable figures.

A precursory glance at Fig. 3 indicates that there are substantial differences in terms of weight and style. In light of the methodological principle that larger teams need heavier methodologies (Cockburn 2007) it perhaps comes as no surprise that of our core methodologies in this work, the heaviest, DSDM, has generally speaking the highest parameter values. These reflect not only the audiences that each methodology wishes to address but also the historical roots from whence they came. For example one major difference between XP and DSDM is methodological scope. DSDM embraces a wider set of concerns than mere product development and offers advice on a broad range of topics including quality management, risk management and configuration management. This contrasts with the sharp XP focus on defining a tightly woven set of complementary and reinforcing practices centered around the production of code and the management of change.

Agile Charting

The iterative nature of agile projects is perhaps best captured by the notion of *process cycles* (Cockburn 2007), which generalizes an idea first found on (Wells 2009a) and which we extend to *agile charting*. Agile charts should be read in such a manner that each cycle of an outer circle implies multiple rotations of its inner circle. Thus in Fig. 4, which illustrates a generic agile process, each increment implies the traversal of one or more iterations which in turn involves the passage of one or more days. During each cycle the same tasks are repeated thereby conveying the cyclical nature of the process. Template agile charts may well exist at the enterprise level but they are intended to be tailored at the project level to reflect their specific vagaries.

For reasons of simplicity a number of activities that ordinarily belong to pre- and post-project phases are omitted from Fig. 4. These include not only the formulation of the business case (known as "system metaphor" in XP or "product vision" in Scrum) but also the benefits review and other ancillary activities. Should the project context be sufficiently important then an additional outer cycle could be added (or other visual metaphor used such as the one presented in Fig. 6 later in this chapter) though this is often omitted. Note that the precise placement of activities on the cycles does not have any temporal implications (e.g., testing in Fig. 4 must not necessarily occur after the cycle halfway mark).

Agile charts are a powerful management and communication instrument and can be used from the outset of a project to clarify when specific activities occur (e.g., stand-up meeting should take place at the start of each day or retrospective workshops are expected to occur at the end of each iteration). Agile charts also make evident any

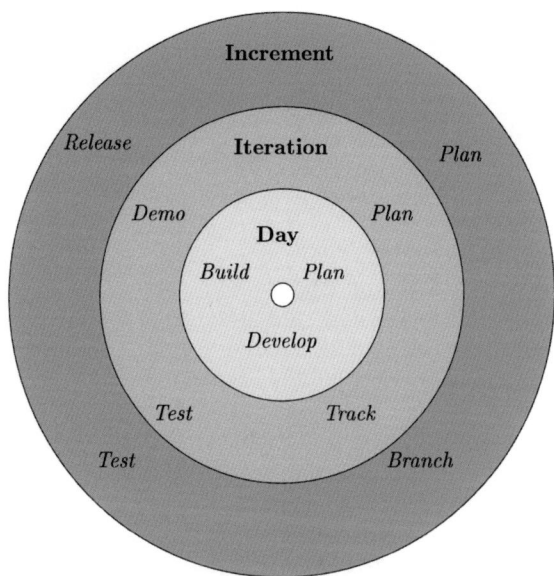

modifications or departures from the standard that one might wish to undertake in
an agile process. For example we shall annotate our process cycles to indicate where
we believe risk activities should occur within the process. Of course, this does not
preclude other amendments to the chart for reasons other than risk management (e.g.,
production of artefacts, position of quality gates). Agile charting is a practice that
should be understood both in methodological and project specific terms and is usually
a combination of both. This means that whilst some activities reflect the practices
of a chosen agile methodology (e.g., daily Scrums) other activities might reflect the
specific environment of a project (e.g., the decision to deploy a nightly build asset
into a centralized test environment). Rather surprisingly, there is little reference in the
literature to the underlying concept of process cycles and few mappings of common
methodologies can be found. We aim to rectify, in part, this deficiency by proposing
agile charts for those methodologies under examination in this work.

Further developing this notion, Fig. 5 illustrates some additional features of agile
charting which might be useful. The chart on the left illustrates a "slice" along which
recurrent activities (e.g., testing, quality or deployment) are placed. For example,
we could slice testing to frame the question of when precisely each form of testing
should occur (e.g., automated unit testing on a daily basis as opposed to iteration
cycle integration, performance and load testing). For example, the team might feel it
necessary to consider shifting some iteration testing to the daily level (e.g., performance testing) if this were appropriate given the circumstances of the project (e.g.,
the need to deliver a high performance product). In this way, agile charting simpli-
fies both the framing of questions and the communication of decisions. The ability
to define slices relies on the recursive nature of activities. However, care should be
taken not to engineer the placement of these activities just so that they lie in a specific

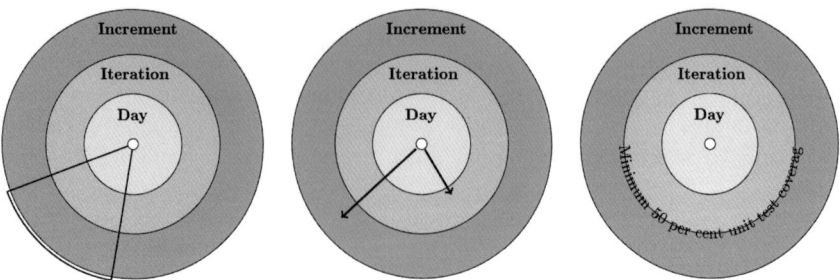

Fig. 5 Variations on Agile Charting. Published with kind permission of © Alan Moran 2013. All Rights Reserved

slice (instead just draw a bubble around the related activities) since to do so would be to distort the original intent of the chart.

The middle of Fig. 5 superimposes a clock face with iteration and increment hands that give a temporal indication of where the project is. This is an example of drafting as described in (Cockburn 2007) and permits anyone to get an instant impression of the project stage in a non-intrusive manner (e.g., without having to bother team members). Agile chart clock faces can be set during the daily stand-up meetings but require that the placement of activities on chart does in fact accurately reflects the project's true dynamic structure (e.g., the use of a clock based on a generic agile chart can be misleading and might even be unhelpful).

A final example of what can be communicated with an agile chart is shown on the right hand side of Fig. 5 and concerns the basis on which the decision to move from iterative development to incremental deployment is taken. We refer to this as the "escape velocity" and define it as the point during product development (at the iteration level) at which the quality and related criteria have been fulfilled thus enabling an increment to be released. For example, a set of development metrics might be defined along with targets that must be met (e.g., at least 50 % unit test code coverage, tangle index of less than 5 %). Should these not be reached during an iteration then at least one further iteration is required during which time new features may be developed alongside the pursuit of these targets. Only when the escape velocity has been acquired (e.g., 55 % unit test code coverage, 4.5 % tangle index) may an increment be considered. Thereafter it behoves the team to maintain these targets (i.e., remain "in-flight") and try to avoid falling back under the escape velocity. Care should be taken not to confuse this concept with feature driven decision criteria that are derived from quality attributes or definitions of done—for this there is already ample accommodation (e.g., MoSCoW).

State of Agility

One the the most comprehensive surveys of agility reports annually on the state of affairs of the discipline and the following statistics taken from a survey of over four thousand respondents (Versionone 2012) reflect (in a qualified manner) the current state of affairs. It also offers important insights for managers responsible for promoting and sustaining agility within their organisations.

- Companies tend not to employ agile project management exclusively though over a third of respondents make the claim (on behalf of their companies) that at least three quarters of their projects do so. The veracity of these claims is, however, unclear given that less than one fifth of respondents occupy leadership positions in their organisations.
- The champions of agility in organisations appear overwhelmingly to be executives or managers with just under one fifth of Scrum Masters apparently holding such roles hinting at small firms where managers assume operational activities. Somewhat incongruently about a third of respondents cited (lack of) management support as a barrier to the adoption of agility.
- Most agile teams are co-located with approximately one third of respondents having had experience of distributed teams.
- Under methodologies Scrum is the clear market leader with just over half claiming to use it and an additional ten percent using some form of Scrum variant (often referred to Scrum-but[9]). XP, in its purist form, is employed by very few organisation though many more admit to using it in conjunction with Scrum, albeit precisely how this is to be understood is not clear from the survey. Finally DSDM languishes together with FDD and agile UP towards the bottom of the scale.
- Techniques such as daily stand-ups, iteration and release planning, unit testing and retrospectives together with continuous integration, automated builds and burn-down charting continue to constitute the core practices of agility. This is reflected strongly in the popularity of tools for which bug trackers, automated build tools, wikis, taskboards and unit testing tools top the list. Amongst cited tools are the usual suspects (e.g., Microsoft Office, JIRA) together with a host of tools ordinarily found in traditional forms of software development not related to agility (e.g., Clear Case, Rational).
- Failure of agile projects is attributed by respondents to tensions between agility and corporate culture including pressure to adhere to Waterfall practices. There is perhaps a self-serving bias to be expected in these responses since a very high proportion of respondents identify very strongly with agility and might be less likely to criticize it.
- More mature forms of agility, such as agile portfolio management, do not appear to be established or popular. Unfortunately data concerning the distribution of sizes of companies is not available as this may well be an explanatory factor.

---

[9] Though this term is often applied derogatorily by purists, it accurately reflects the reality that Scrum is often adapted to the circumstances of an enterprise.

- It seems old habits die hard as the greatest concerns about adopting agility revolve around lack of planning, loss of management control (and thus lack of support), lack of predictability and discipline concerns. Such topics are tackled in (Boehm and Turner 2009; Larman 2003) though some methodologies (e.g., DSDM) tackle these matters directly.
- Perception of the benefits of agility focus on time-to-market, ability to deal with changing priorities, better IT/Business alignment and higher productivity. Risk features further down the list (in spite of which over three quarters of respondents claimed that management of risk improved) and value is never explicitly mentioned though perhaps implicitly implied.
- Support for agility is expressed in terms of training and the presence of consultants (bearing in mind that approximately one fifth of respondent were themselves consultants!).

The results of this survey should be treated with a little caution since they draw primarily on the experiences of practitioners from North America and Europe who comprise nearly 90 % of respondents, almost half of whom have relatively little experience (i.e., less than 2 years) of agility.

Other rich sources of survey results exist, some of which also include raw data, suggesting that agile and iterative project approaches exhibit high performance when measured on the basis of quality, value, financials (e.g., return on investment) and on-time delivery (Ambler 2013d). Interestingly, one such survey, (Ambler 2013a), gives broad indications concerning the manner in which agility is deployed citing examples of very large team sizes (e.g., more the fifty team members). It also notes that teams are invariably dispersed though co-located teams enjoy higher success rates. This reflects comments found in (Leffingwell 2007) that at scale geodispersion is a fact of life. There are also indications of the circumstances that enable agile teams to be truly successful (e.g., agile skill levels, presence of agile champions) with the survey results being confirmed by multiple other sources (Lines and Ambler 2012; Leffingwell 2007).

Management Perspective

The literature concerning agile management appears to focus exclusively on matters of operational management and leadership, citing notions such as empowerment, emotional intelligence and collectivism and contrasting these with command and control approaches. Other studies approach agility in a survey like manner describing and comparing various methodologies (Larman 2003; Augustine 2008; Anderson 2003). There is sadly little commentary on the governance of agile project management or its relationship to strategic business or corporate management and accordingly topics such as agile portfolio management and strategic alignment receive only limited attention (Krebs 2008). We believe that agility ought to be understood as a core competency and an enabler of competitive advantage and thus rightly belongs

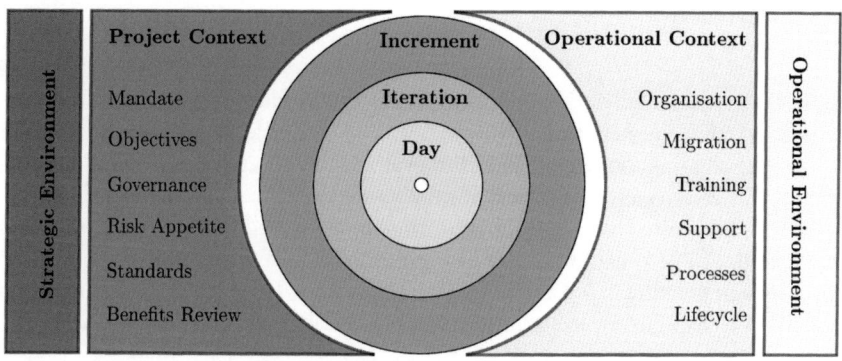

Fig. 6 Strategic and Operational Context of Agility. Published with kind permission of © Alan Moran 2013. All Rights Reserved

under the remit of strategic management. Indeed in the final chapter we will cite the example of salesforce.com which undertook a radical restructuring of management, based on an adapted form of Scrum, to revitalize an organisation already straining under the weight of bureaucratic controls.

Figure 6 depicts agile process cycles in the context of the strategic and operational environments. Projects arise in the context of a strategic portfolio[10] which has overall responsibility for governance (incl. risk appetite and attitude), standards (e.g., risk management principles) and support (e.g., project office functions). It is at the strategic level that matters of agile dynamic capabilities and competencies along with the strategic advantage they pose find expression whereas at the portfolio, programme and project levels[11] this strategy governs, directs and is itself transformed into implementation. Key to a smooth transition is this alignment of dynamic strategy and agile capability which must ensure that tensions do not arise (e.g., the imposition of a plan-driven strategic management approach). Some methodologies (e.g., DSDM) appear to be well suited for this task as they are already equipped with the necessary project and managerial infrastructure.

Thus a project acts as a container for agile activity within which the terms of reference are set (e.g., mandate, governance, standards). However, projects are, by their very definition, temporary organisational constructs that are outlived by the deliverables they produce. Therefore agile capability must be transferred into the operational context which sets new terms of reference (e.g., product release and life-cycle management) and introduces new activities with which agility must engage (e.g., migration, support). It is in this context that we find notions of agile infrastructure and continuous delivery enshrined in the practices of DevOps (Loukides 2012). Thus an agile methodology must bridge both worlds if it is to remain capable of sustaining long term value creation and preservation. This remains a major organisational challenge which requires considerable skill to master.

[10] A further subdivision at the programme level may also apply for related projects that might benefit from such an organization.

[11] Designated as "project context" in Fig. 6 for reasons of simplicity.

Concluding Remarks

Agility is fundamentally driven by principles in which transparency, open communication and empowerment thrive. We capture the iterative and incremental character in the agile charting tool which we will repeatedly use throughout this book and recommend its creative use as a means of framing issues and communicating decisions, activities and goals. The methodologies that we introduced in this chapter illustrate the breadth of interpretation in this discipline and whilst we recognise the product development thinking in XP and Scrum we also acknowledge the maturity and depth of DSDM as an agile project management methodology that reaches out to a more enterprise notion of agility.

Project Risk Management

Abstract We start out by defining what we mean by risk in the project context and explore cultural attitudes towards risk which constitute an important element of agile risk management. From there we review a number of risk management frameworks that span the domains of project and enterprise risk management distilling their similarities into a generic risk management model, which we use to explain the key activities involved in identifying, assessing, treating and monitoring risks. Finally we close on some comments regarding enterprise risk management and how the corporate view of risk relates to the project risk management. Our discussion is indicative of the state of affairs of risk management today and provides the language, process and tools for discussing agile risk management in the Chapter "Agile Risk Management".

Definition of Risk

Throughout this book we understand risk in the context of projects and define it as *uncertainty that has an impact on project objectives* (Hillson 2009). Risk is commonly described in terms of its components of *likelihood* and *impact* though we shall refer to these collectively as the *risk exposure*. Note that in the literature it is not uncommon to use the terms risk and risk exposure interchangeably. For example a statement such as "the risk of at least one financial loss of more than one million dollars within the next six months" incorporates both impact and frequency, a common proxy for likelihood, as well as proximity. Risk exposure is sometimes formally expressed as a co-ordinate pair, or the individual components may be converted into numbers and the exposure computed as their product. Particular care should be taken, however, when calculating exposure in this fashion, as converted numerical values may in fact be nothing more than ordinals for which their mathematical product carries no semantic meaning whatsoever, though its magnitude may nonetheless have some indicative character about it. This is commonly the case when using T-shirt sizing to rate risks with a scoring to convert them into values (e.g., "S", "M" and "L"

A. Moran, *Agile Risk Management*, SpringerBriefs in Computer Science,
DOI: 10.1007/978-3-319-05008-9_2, © The Author(s) 2014

may be assigned a values of 1, 2 and 4 respectively). We shall return to this point later in the Chapter "Agile Risk Management" when we propose *risk scoring* as a means of converting exposure directly to a numerical scale for the purposes of monitoring and tracking.

Risk is often equated with uncertainty. Uncertainty may or may not be something that we believe we can accurately measure and thus we chose not to make the differentiation between "risk" (knowable in probabilistic terms) and "uncertainty" (unknowable randomness) as suggested by Knight (1921). Later when we address risk assessment we will, however, argue that estimation of uncertainty is inherently a subjective matter and since we will rarely have recourse to strictly probabilistic means of judging risk, we consider this distinction to be moot for all practical purposes. It is important to understand though that not all forms of uncertainty are relevant and that we understand risk in this context to be related to uncertainty that has a bearing on our objectives. To employ an analogy, the outcome of a horse race is uncertain but irrelevant unless we have placed a bet on a horse, in which case the uncertainty directly affects (in a positive or negative fashion) our objective of making money.

Risk can have upside (positive) as well as downside (negative) consequences. It is commonly the case that negative risk is implicitly implied and this appears to be the prevailing view in the agile literature. However, opportunity (positive risk) can manifest itself in many forms ranging from operational issues (e.g., the need to cater for higher than anticipated demand for a service) to strategic matters (e.g., extensibility of product into new market segments). For example, consider the opportunity in a software product that its use may exceed expectations by tenfold. Since this is a desirable outcome with considerable financial upside, one or more measures might be considered to try to promote this result. These might range from technical measures (e.g., improving the algorithms to ensure higher scalability), to infrastructural amendments (e.g., switching to a load balancing topology) and post-deployment actions such as increasing awareness of the product (e.g., marketing). Were all enacted, it would be likely that the risk would be realised and the lost opportunity of still further use of the product would be reduced as the product begins to saturate the market.

The definition of risk as used in this book is consistent with that found in Enterprise Risk Management (Moeller 2011; AS/NZS 2004; ISO 2009) and related practices (HM Treasury 2004; IRM et al. 2002; Protiviti 2013) including its application to IT (ISACA 2009a, b) which we shall return to later. Suffice it to say for now that project risk ought not acquire an "island mentality" and should be managed in the context of wider enterprise goals, consistent with their view on risk attitude, appetite and tolerances.

Cultural Attitudes to Risk

Risk assessment is not an entirely rational stochastic activity, wherein data is precise and models accurately reflect the realities of the risk scenario. Indeed, our experience suggests that this is seldom, if ever, the case but rather that assessments are largely a subjective and even visceral affair. Accordingly it is worth reviewing the literature

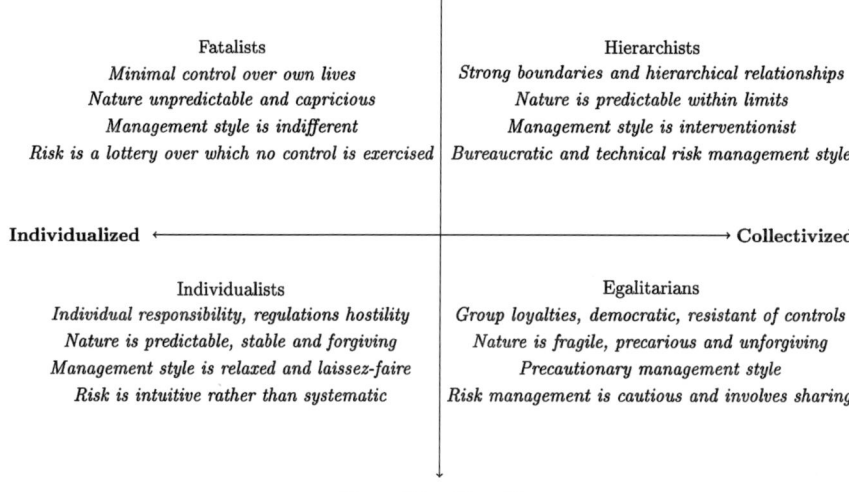

Prescribing Inequality

Fatalists	Hierarchists
Minimal control over own lives	*Strong boundaries and hierarchical relationships*
Nature unpredictable and capricious	*Nature is predictable within limits*
Management style is indifferent	*Management style is interventionist*
Risk is a lottery over which no control is exercised	*Bureaucratic and technical risk management style*

Individualized ←————————————————————→ **Collectivized**

Individualists	Egalitarians
Individual responsibility, regulations hostility	*Group loyalties, democratic, resistant of controls*
Nature is predictable, stable and forgiving	*Nature is fragile, precarious and unforgiving*
Management style is relaxed and laissez-faire	*Precautionary management style*
Risk is intuitive rather than systematic	*Risk management is cautious and involves sharing*

Prescribing Equality

Fig. 7 Cultural theory risk-taking typology. Adapted from (Adams 1995). Published with kind permission of © Alan Moran 2013. All Rights Reserved

to discover what it has to say about the social and cultural attitudes towards risk assessment and the perceptions of risk and reward.

We begin with the view commonly referred to as "cultural theory" as espoused by Thompson et al. (1990) that where scientific rationalism proves insufficient for the assessment of risk we are guided by assumptions, inference and beliefs. Ultimately risk should be understood as a cultural construct that admits a plurality of rationalities, giving rise to differences in opinion born of the variances in premises from which arguments are made. The origins of this theory can be traced to a typology of attitudes proposed by Holling (1979) and later developed by Schwarz and Thompson (1990) wherein decision making patterns of managers faced with incomplete information appeared to rely on their beliefs about nature which in turn influence their management styles. This led to a classification of risk-taking patterns described in Adams (1995) and summarized in Fig. 7 which includes the anthropocentric traits cited in the original typology.

The metaphor of a "risk thermostat" is discussed in Adams (1995) to illustrate the notion that everyone has a propensity to take risks that are influenced by their perception of risks and rewards. Ultimately management of risk becomes a balancing act that is influenced by cultural premises that underpin reasoning in relation to risk assessment. Indeed apparently irreconcilable conflicts over how risks are judged may, according to this theory, be attributable more to the premise on which they are based than the rationality of their arguments. Admittedly "empirical evidence for this theory is ... sparse" (Adams 1995, p. 38) and its application to account for different styles of balancing risk can be speculative.

An alternative classification is presented in Hillson (2009) which presents the following list of risk attitudes.

- *Risk-averse*: Preference of secure payoffs, common sense and facts over theories. Propensity to over-react to threats and under-react to opportunities.
- *Risk-seeking*: Preference towards speculation and unafraid to take action. Propensity to underestimate threats and overestimate opportunities.
- *Risk-tolerance*: Indifference towards uncertainty that lends itself to reactive rather than proactive measures. Propensity to fail to appreciate importance of threats and opportunities alike.
- *Risk-neutral*: Impartial attitude towards risk and act in the interests of significant benefits. Propensity to focus on the longer term.

These observations are based on Murray-Webster and Hillson (2008) in which the authors propose that risk attitude is influenced by the "triple strand" of conscious factors (e.g., visible and measurable characteristics), subconscious factors (e.g., heuristics) and affective factors (e.g., visceral feelings). Their research appears to indicate that it is at the point of choosing a risk response that risk attitude plays a role (i.e., tendency to engage in a risk or withdraw from it) and suggest a simple model for managing risk attitudes within a group. The relevance of this research for agile projects is the appreciation that different members within a team may hold fundamentally differing views towards risk and that the conflict that arises when assessing risk ought not be judged entirely in terms of the rationality of opposing arguments. The emotionally intelligent agile project manager ought, therefore, to invest some time understanding the risk disposition of the members of his/her team.

Commenting on organisational and national culture, Hofstede (2003) described an "Uncertainty Avoidance" Index (UAI) against which it became possible to score nations.[1] Uncertainty avoidance is defined as "extent to which the members of a culture feel threatened by ambiguous or unknown situations and have created beliefs and institutions that try to avoid these" (Hofstede 2013) which goes on to describe the index in greater detail.

> The uncertainty avoidance dimension expresses the degree to which the members of a society feel uncomfortable with uncertainty and ambiguity. The fundamental issue here is how a society deals with the fact that the future can never be known: should we try to control the future or just let it happen? Countries exhibiting strong UAI maintain rigid codes of belief and behaviour and are intolerant of unorthodox behaviour and ideas. Weak UAI societies maintain a more relaxed attitude in which practice counts more than principles (Hofstede 2013).

Just how widely UAI varies is illustrated in Fig. 8 which uses data sourced from Hofstede (2013), which was collected between 1967 and 1973, to depict the attitudes towards uncertainty amongst the G-20 of major economies.[2]

---

[1] UAI is merely one of several cultural dimensions including power distance, individualism versus collectivism, masculinity versus femininity, long versus short termism and indulgence versus restraint.

[2] G-20 comprises of nineteen member states plus the European Union which is excluded from Fig. 8.

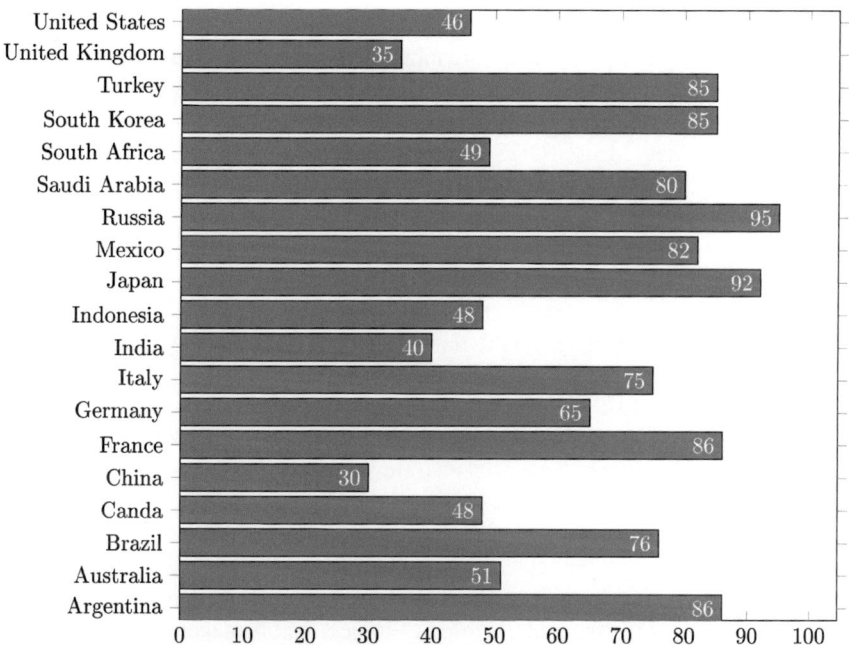

Fig. 8 Hofstede uncertainty avoidance index for the G-20 member states. Published with kind permission of © Alan Moran 2013. All Rights Reserved

Since uncertainty avoidance is related to risk attitude it has a bearing on how individuals may perceive and attempt to control risk within projects and organisations. In particular in the light of risk classification and its influence on choice of risk responses it could be argued that risk propensity is the dominant characteristic in risk behaviour, a view that is also echoed in Sitkin and Pablo (1992). These considerations become particularly acute when dealing with teams distributed across several cultures which is likely to be the case in geodispersed agile teams that are commonly featured in enterprise forms of agility (e.g., DSDM, DAD and SAFe).

Synthesis of Risk Models

There is a considerable degree of consensus amongst the common risk management frameworks such as the Management of Risk (M_o_R®) of the British Cabinet Office[3] (Office of Government Commerce 2007), the Orange Book of the UK HM Treasury (2004), the IRM, AIRMIC and ALARM Standard (2002) and the Practice Standard for Project Risk Management of the Project Management Institute (2009).

---

[3] Formerly this function was assumed by the Office of Government Commerce.

The Management of Risk framework, sponsored by the British Cabinet Office, is founded on principles that reinforce organizational context and stakeholder involvement in the context of a structured and directed risk management effort. It is tightly integrated into other UK Government frameworks such as Managing Successful Programmes, MSP®(OGC 2011b), Projects in a Controlled Environment, PRINCE2®(OGC 2009) and the IT Infrastructure Library, ITIL®(OGC 2011a) and accordingly may appeal to mature process oriented organisations. A similar initiative on the part of the UK Government Treasury (2004), covers the same ground and provides advice and guidance in relation to risk management in UK Government departments. This publication discusses the notion of the extended enterprise (e.g., partner organisations) and appears to have a greater appreciation of the influence of enterprise risk management.

The Institute of Risk Management (IRM), Association of Insurance and Risk Managers (AIRMIC) and ALARM The National Forum for Risk Management in the Public Sector together also produced a simple guide to risk management replete with examples, advice and suggestions. Finally the Project Management Institute has augmented its flagship project management methodology with a standard describing the practice of risk management which provides rather detailed information concerning techniques applicable to the project context. An excellent source of additional information concerning the general topic of project risk management can be found in Hillson (2009). This is by no means an exhaustive list but is sufficient to illustrate the convergence of thinking within the risk management community. The similarity between these frameworks permits us to synthesize them into a generic process, the key stages of which are found in Table 2. In turn each of the frameworks can be broadly mapped to these stages as listed in Table 3.

All of the frameworks cited thus far have a generic appeal and are not specific to a particular sector (e.g., IT, Finance). They all emphasize the continuous nature of risk management together with its embedding within a wider organizational context. There remain minor differences in approach, scope and terminology reflecting the needs of the audiences that each framework seeks to address but these will not concern us here. Some of this consistency owes itself to the existence of standardized vocabularies such as ISO (2002) which was later revised in 2009.

We now turn our attention to each of the stages of our synthetic risk management process (described in Table 2) and comment on what is generally understood to occur within each stage. The purpose of this discussion is to set the foundations and expectations for the agile risk management process to be discussed in the Chapter "Agile Risk Management".

Initiation and Planning

Since risk is concerned with relevant uncertainty in relation to project objectives, it is necessary to have these defined before risk management can be initiated. Typically this occurs as part of the formulation of the business case or change mandate of

Table 2 Common stages in risk management frameworks

Stage	Description
Initiation and planning	This stage typically covers a variety of activities including the scoping of risk management and setting of objectives, determination of stakeholders, discovery of organisation and terms of reference, assessment of the risk environment, appetite and tolerances
Risk identification	Activities here usually encompass the preparation of a risk register and determination of risks facing the project through which an understanding of project risk is established (e.g., risk factors). Formulation of risk indicators (e.g., early warning signs) may also occur at this stage
Risk assessment	Assessment of the likelihood and impact of each risk perhaps together with an estimation of their proximity. Techniques may either be quantitative or qualitative as appropriate. There may also be some form of generation of a risk profile to graphically illustrate the contents of the risk register. Reporting might also extend to an aggregate view of risks as a portfolio together with monetized estimates of overall risks
Risk treatment	Determination of appropriate risk response strategies (see Table 6) together with specific countermeasures or contingency plans. This usually includes an assessment of the costs of risk actions which in turn need to be balanced against costs incurred should a risk be realized. The residual risk remaining after the response has been applied should also be assessed and may constitute grounds for further actions
Risk monitoring	Assurance measures to gauge whether or not planned responses have been implemented and whether or not they were effective (along with corrective action where appropriate). This stage caters for accountability of the process and ensures an adequate level of reporting is present (e.g., trends, costs, distribution of risks by category) and that communication regarding risk is appropriate
Risk review	Ordinarily this forms part of a benefits or post-project review and affords the opportunity to assess the performance of the risk management process as a whole. Lessons learned may be gathered and summary information of use to future projects may be compiled

a project during which stakeholders, who are important risk participants, are identified. When we return later to enterprise risk management we will note that the terms of reference for risk management are often determined by the enterprise attitude towards risk (e.g., risk appetite and permissible risk tolerances for projects) which may also include concessions to specific departments within the enterprise where a greater level of risk may be tolerated (e.g., research and development). Indeed several approaches make references to such matters including the risk environment/context and extended enterprise cited by HM Treasury (2004) or the organisation strategic objectives mentioned by IRM et al. (2002). Ultimately the risk character of the project must be made clear to project members from the outset along with the manner in which risk and reward should be balanced. In addition Crowe Horwath (2011) is a detailed source of guidance on risk appetite and tolerance (albeit in the context of enterprise risk management).

Table 3 Comparison of stages of major risk management frameworks

Stage	M_o_R®	IRM/AIRMIC/ ALARM	Orange Book	PMI® BoK
Initiation and planning	Identify context	Organisation's strategic objectives	Risk environment/ context, the extended enterprise	Plan risk management
Risk identification	Identify risks	Risk identification, risk description	Identifying risks	Identify risks
Risk assessment	Assess estimate and evaluate	Risk estimation, risk evaluation	Assessing risks	Perform qualitative risk analysis, perform quantitative risk analysis
Risk treatment	Plan and implement	Risk treatment	Addressing risks	Plan risk responses
Risk monitoring	Communicate	Residual risk reporting, monitoring	Reviewing and reporting risks	Monitor and control project risks
Risk review	Embed and review	(Monitoring)	Communication and learning	(Project review)

The basis of risk communication are also typically laid during the planning stage. This might include expression of risk expectation as embodied in the risk appetite of the organisation but could also encompass information on the risk management capability and how well it is managing risk together with the actual risk levels within the organisation and the extent to the which the project contributes in this manner (ISACA 2009a). Clarity concerning the risk culture, defined in ISACA (2009a) as comprising of behaviour towards risk, behaviour towards following policy (i.e., compliance) and behaviour towards negative outcomes (e.g., blame culture), is also required at this stage. These combined efforts set the tone for risk management and give project members the opportunity to assess where they stand in relation to risk management practices of the organisation.

Risk Identification

This is the first operational stage of the generic process and requires the identification of potential threats and opportunities facing the project. These may be external (where a lesser degree of control might be expected) or internal (where more control might be possible). All frameworks recommend the creation of a risk register, see Table 4 which is indicative of typical project risk attributes, to record and later manage risks though the precise attributes vary according to the scope of the exercise (e.g., project, enterprise). For example IRM et al. (2002) classifies risks according to its own

Table 4 Common risk register attributes

Attribute	Description
Identifier	Code or name identifier with which the risk may be referred to. Unless this is actually used somewhere it ought to be considered optional
Description	Description of the precise nature of the risk. We return to specific techniques applicable to the agile context in the Chapter "Agile Risk Management"
Classification	Placement of the risk in some form of categorization. Typically this is based on risk drivers though the project may choose to derive a classification based on other factors
Likelihood	The probability with which the risk event is likely to occur. Since this, however, is a stochastic concept, frequency (with respect to a specific timeframe) is commonly used as a proxy. This information can only be gathered during the risk analysis stage
Impact	The impact the risk event, should it occur, is likely to have. This information can only be gathered during the risk analysis stage
Owner	The individual (or function) tasked with implementing the risk measure
Response	The intended risk strategy to be adopted in order to manage the risk. This information can only be gathered during the risk management stage
Measure	The measures (if any) to be taken to address the risk. Until enacted this is merely a planned response
Residual likelihood	The resultant likelihood of the risk event once the measure has been implemented
Residual impact	The resultant impact of the risk event once the measure has been implemented

scheme[4] and suggests inclusion of potential actions for improvement and strategy and policy developments whilst Office of Government Commerce (2007) includes rather fine grained attributes and also includes risk proximity.

Risk identification may require the gathering of data and the determination of the main contributing factors of risks (i.e., those that determine the frequency and/or impact of risk events). In some cases risk indicators, which constitute early warning signs in control based systems, may also be formulated and assigned to risks at this stage. Generally speaking brainstorming workshops, checklists, interviews soliciting expert opinion are the most common techniques applied to identify risks.

Risk Assessment

Risk assessment involves the decomposition of risks into the components of likelihood and impact (though some frameworks also add proximity) either by applying subjective judgement (qualitative) or use of a numerical model or scales (quantitative). A finer classification, presented in Table 5, is perhaps more indicative

---

[4] This scheme comprises of strategic, operational, financial, knowledge management and compliance.

Table 5 Methods for assessing risk

Method	Description	Remarks
Expert opinion	Solicitation of views from those deemed to be most knowledgeable	Practical approach that relies on the identification of appropriate individuals and a means of resolving differences of opinion (e.g., Wideband Delphi)
Data diving	Use of data to assess impact and frequency over a specified time period as a proxy for likelihood	Useful when past data is available, reliable and is indicative of future behaviour. Risk of transference when applied to innovative projects where no prior experience exists
Stochastic	Application of statistical models (e.g., Monte Carlo simulation) that capture the essentials of the underlying behaviour	Appropriate when data is not available but circumstances lend themselves to accurate a priori modelling. Requires considerable theoretical skills and techniques may vary according to which component is being assessed
Cultural	Reliance on the inherent subjective assessment of both the nature of events and their likelihood and impact	Admits a plurality of rationalities into the assessment of risk

of how assessment is really conducted and reflects the reality that multiple approaches may be adopted simultaneously.

Each of the approaches mentioned in Table 5 have their strengths and weaknesses. For example a data driven approach might be appropriate for a IT network project within a highly controlled and regulated environment typical of an enterprise with a mature ITSM[5] culture. In such a situation the failure of a network component is likely to be noted by an event monitoring system which in turn might record lack of availability as part of an integrated incident management process that may be linked to service level management. In this context we may feel some degree of confidence in the data we receive and be willing to rely on them for our estimates. Moreover such estimates subsume a multitude of causes (e.g., hardware failure, misconfiguration) that would otherwise be hard to specify and trace in an analytical model. On the other hand, were the project to employ novel technologies of which the organisation has no prior experience then it would become appropriate to augment these estimates with a stochastic model that predicts likely outcomes perhaps based on component estimates provided by suppliers. In practice most risk assessments rely on brainstorming workshops and the input of experts alone and for most purposes this is entirely adequate.

Traditional approaches to risk management create a *risk map* by plotting frequency against impact and then define linear contours of constant risk exposure in order to form partitions for the purpose of risk management as in Fig. 9. Thus the following

---

[5] IT Service Management (ITSM) is a process oriented approach to the management and delivery of IT as a service and commonly employs frameworks such as the IT Infrastructure Library (ITIL) or Control Objectives for Information and Related Technology (COBIT).

Fig. 9 Traditional risk map.
Published with kind permis-
sion of © Alan Moran 2013.
All Rights Reserved

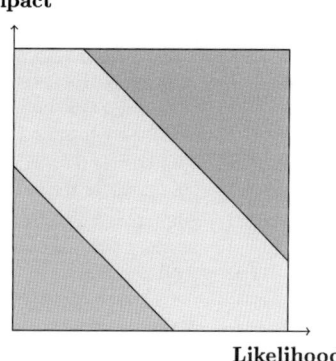

Impact

Likelihood

diagram highlighting areas of high (top right band), medium (middle band) and low (bottom left band) risk is commonplace. High risks require immediate and urgent treatment, medium risks can be tolerated for a period of time but will eventually require management and low risks can be ignored to the extent that controls can be loosened (or at most contingency plans prepared). This technique is promoted in several places including ISACA (2009b) where guidance is provided on using such graphs to aggregate individual risks into composite ones.

We return to this representation to comment that in plotting the magnitude of impact this graph fails to distinguish between positive and negative risk and suggest that an alternative representation might be more appropriate in the agile context given the willingness to embrace change in the pursuit of reward. This is not to say that such representations do not already exist (for example a "double probability-impact matrix" is described in Hillson (2009) and elsewhere). Once assessed, the estimates should be entered into the risk register and reviewed should circumstances change significantly (e.g., revision of inputs from experts, perception of increasing or decreasing external risk that might impact already formulated risks).

Risk Treatment

The treatment of risk begins with the identification of suitable risk response strategies based on the classical list cited by most frameworks (see Table 6). Until relatively recently exploitation of risk seldom featured in this list.[6] Indeed some even describe the notion of exploiting risks (i.e., taking opportunities) as not belonging to one of the key risk measures (or "internal controls") but rather "it is an option which should be considered whenever tolerating, transferring or treating a risk" (HM Treasury 2004, p. 28).

---

[6] Acronyms such as TARA (Transfer, Accept, Reduce, Avoid) and SARA (Share, Accept, Reduce, Avoid) were commonplace.

Table 6 Classical risk response strategies

Risk strategy	Description
Accept	Appropriate for low risk exposure. No specific action will be undertaken to mitigate or manage the risk. Instead a contingency plan will be developed to tackle the eventuality that the risk may be realized
Reduce	Appropriate for medium negative risk. Actions will be undertaken to reduce either the impact or the likelihood of occurrence of the event
Exploit	Appropriate for medium to high positive risk. Measures will be undertaken to increase either the frequency or the impact of the event
Share	Appropriate for low frequency high impact positive risk where the efforts to manage the risk alone may not be warranted owing to the relative unlikelihood of it occuring. Measures will be undertaken (with others) to share the risks and rewards
Transfer	Appropriate for low frequency high impact negative risk where efforts to contain it may be beyond the capabilities of the team. Measures include the transfer of the risk to a partner capable and willing to treat it (e.g., out-sourcing to a specialist)
Avoid	Appropriate for high negative risk. The activities that give rise to the risk will not be undertaken

Table 7 Classical controls for negative risks

Risk control	Description
Preventative	The most common form of control that seeks to limit the possibility of a threat being realized (e.g., segregation of duties, authorization, encryption)
Corrective	Measures to address threats that have already been realized and thus these have a contingency nature about them (e.g., recovery or restoration activities)
Directive	Requirements that ensure or at least encourage behaviour that avoids the threat (e.g., security policy)
Detective	Controls that make known that threats have been realized (e.g., audit logs, monitoring)

One generally seeks to reduce or avoid negative risk and if this is not possible to transfer it. Not surprisingly both Office of Government Commerce (2007) and HM Treasury (2004) are in agreement about the nature of controls in relation to negative risks (i.e., threats) with both even citing very similar examples. The controls they cite for negative risk are listed in Table 7 and are perhaps familiar to IT security specialists who also employ additional controls such as deterrent activities (e.g., threat of legal action) not listed in this table.

On the other hand, the risk response strategies that apply to opportunities are exploitation or sharing. The treatment of positive risk, like that of negative risk, reduces its residual risk. For example by making a software application more scalable to cater for the risk of increased demand, this risk is reduced in the sense that there is less opportunity for the scaled application to exploit (or to phrase it negatively the opportunity to exceed the capabilities of the software is now less than it was before). Sometimes it is infeasible or uneconomical to treat a risk and we must simple accept

Impact

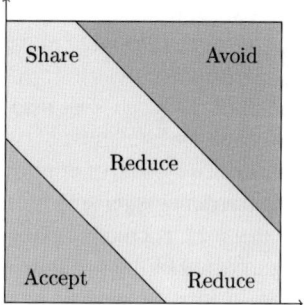

Fig. 10 Traditional mapping of risk responses strategies. Published with kind permission of © Alan Moran 2013. All Rights Reserved

that it might occur. The best we can do is to set aside contingency plans for this eventuality.

Often risk exposure components are used to map risks to potential response strategies as advocated in Collier (2009) which cites COSO (2004) which is typically done using a risk map (see the mapping in Fig. 10). We shall return again in the Chapter "Agile Risk Management" with our own interpretation of risk response strategies and their mapping using risk exposures. In any event, the chosen measure to tackle a risk will inevitably include an assessment of its costs, which need to be balanced against costs incurred should a risk be realized. Note that the costs of proactive risk measures are ordinarily embedded in normal project activity costs whereas reactive risk treatment must be set aside in the form of a risk budget for the eventuality that such risks are in fact realised.

The effectiveness of a proposed measure can be assessed in terms of the residual that remains once the measure has been enacted. This is referred to as the *residual risk* and details of such risks together with the original risk actions need to be recorded in the risk register. Furthermore, new risks may come to light as a direct result of implementing a risk response. Known as secondary risks, these too should be recorded in the risk register[7] and managed appropriately.

Risk Monitoring

Risk monitoring involves assessing whether or not the risk profile of the project has altered (i.e., it measures the extent to which risks are being effectively managed through the reduction of inherent risks to residual risks). In addition the tracking of costs and status of the risk budget together with the promotion of risk awareness and communication may also be embodied in this stage of the process. If possible, it is

---

[7] Some registers include an attribute to link secondary risks to the primary risks that gave rise to them.

worth considering indicators of impending realisation of risks and noting them in the risk list for monitoring. This is a topic that arises frequently in risk management and is covered extensively in the literature (ISACA 2009a, b).

Risk monitoring is perhaps best performed by maintaining a risk profile (i.e., recording of risks on a risk map) and tracking key risk indicators which are capable of measuring the likelihood of risks exceeding predefined tolerances. Note that the use of indicators is typical of control based risk management and features somewhat less in project risk management. Of importance is the state of overall project risk and any trends that may be emerging (e.g., increase in the number of risks drifting towards the top right region of the risk map). The review of the risk register confirming accuracy (e.g., inclusion of new risks) and relevance also constitutes part of the monitoring. Indeed it is monitoring that affirms its cyclical and iterative character and ensures the continuous nature of risk management or in the words of Hillson (2009, p. 47) "for risk management, standing still is going backwards". When we discuss this topic in the Chapter "Agile Risk Management" we will adopt a different approach to risk monitoring by using a risk modified Kanban board together with a risk burndown chart which together ensure high visibility of risks as well as their distribution across the project.

Risk Review

Risk review is primarily about providing assurance that the risk management process itself is effective and to ascertain the level of compliance (HM Treasury 2004; ISACA 2009a, b; Office of Government Commerce 2007). Often this takes place during the post project review phase (though later when we address agile risk management we will propose that this aspect ought to be covered during iteration retrospectives). The Management of Risk framework devotes considerable attention to this topic and covers a range of issues including success factors, management support and commitment, development of awareness and continual improvement (Office of Government Commerce 2007). Similarly ISACA (2009a) dedicates several processes to the theme of risk governance which aims to secure an optimal risk-adjusted return and spans a wide range of strategic and operational activities. Lessons learned may be gathered and summary information of use to future projects may be compiled.

Enterprise Risk Management

Enterprise Risk Management (ERM) extends the traditional risk management focus which is chiefly concerned with specific domains (e.g., projects, IT, security, health and safety) to a more holistic view encompassing the entire organisation. The primary frameworks in this regard include AS/NZS (2004), ISO (2009), COSO (2004) and specifically in relation to IT (ISACA 2009a, b).

The scope of ERM supports alignment of risk management practices when tackling complex objectives that are not under the remit of a single project team or department (e.g., mergers and acquisitions, large scale IT migrations, IT security and compliance). It can also support the use of a common risk language throughout the organization (e.g., common scales to assess risk impact) thereby making localized risks more comparable and enabling the aggregation of risks to better understand risk distribution. Indeed ISACA (2009b) provides specific guidance in this respect.

That ERM seeks to improve the "the management of increasing risk mitigation costs and the success rate of achieving business objectives" (Protiviti 2013, p. 3) by helping reduce unwanted performance variability reflects the widely held perception that good governance leads to better performance and that responsibility for risk lies with the board of directors (Connelly et al. 2010). Indeed the board has a duty to achieve an understanding of risk and communicate its significance throughout the organisation. Risk management is repeatedly cited as a central component of governance that features in a variety of IT frameworks (OGC 2011a; ISACA 2009a, 2012a, b). As a result well risk managed organisations better understand the balance on the one hand between risk and reward, and on the other the capabilities of the organisation and its ability to absorb losses or fail to capitalize on opportunities. This in turn builds investor and stakeholder trust who retain confidence in the organisation to weather bad times and cope with changes in the future.

To better understand the nature of ERM it helps to take a look at the financial statements of major listed companies wherein a description of the potential impact of risks on strategic objectives can be found. For example in 2012 IBM noted the threat in the downturn of the economic environment and corporate IT spending budgets (IBM 2012). It also hinted that potential shortcomings in its internal transformation programme may result in it failing to meet its growth and productivity objectives as a result of which competitiveness might suffer and cited additional potential difficulties in innovation that were threatening the reputation and brand of IBM. Far from unsettling investors such frank admissions make clear that these matters have the full attention of senior management and that measures are being considered to address them. For example elsewhere in its filings IBM, which spends approximately $35 billion annually on its supply chain, describes the benefits of risk related activities as follows

> Continuous improvements to supply chain resiliency against marketplace changes and risks have been particularly valuable in maintaining continuity during natural disasters and other disruptive events (IBM 2012).

and goes on to state that it continues to manage and fine tune its supply chain through the application of analytics to risk management. These statements project the image of a mature risk focused organisation that has a clear understanding of where the balance between risk and reward lies.

In light of financial scandals and corruption in the US dating back to the mid-1970s the Committee of Sponsoring Organizations of the Treadway Commission (COSO), a joint venture comprising of the Institute of Management Accountants (IMA), the American Accounting Association (AAA), the American Institute of Certified Public Accountants (AICPA), the Institute of Internal Auditors (IIA) and the Financial Executives International (FEI), set about defining standards covering

Table 8 Comparison of enterprise and project risk management

Enterprise risk management	Project risk management
Wide scope that requires the commitment of senior management and the engagement of the entire organisation	Narrow scope that chiefly concerns the project team and its stakeholders through direction by senior management may occur
Strategic focus on the objectives of the organisation	Tactical focus on project objectives
Concern related to the tangible and intangible assets of an enterprise that underpin its business model	Concern is limited to the project as a change vehicle
Governance function with responsibility for oversight of the management of risk that faces an organisation	Governance restricted to the level of risk management process (e.g., effectiveness, compliance) in relation to project objectives

governance, ethics, risk management and financial reporting. Further scandals involving Enron and Worldcom resulted in the passing of the Sarbanes-Oxley (SOX) Act in 2002 which was followed in 2004 by the COSO ERM Integrated Framework standard (2004) compliance with which was generally accepted to be in accordance with the new SOX legislation. The following broad and comprehensive definition of Enterprise Risk Management (ERM) describes it as,

> a process, effected by an entity's board of directors, management and other personnel, applied in strategy-setting and across the enterprise, designed to identify potential events that may affect the entity, and manage risk to be within its risk appetite, to provide reasonable assurance regarding the achievement of entity objectives (COSO 2004).

We will not dwell further on the topic of ERM so suffice it to say that there exists at the corporate level a mature understanding of risk in the context of which project risk management must find itself. This parallels the relationship between agility and strategic management alluded to already and suggests that concepts such as agility, risk and strategy ought not to be understood as incompatible terms but rather find expression at many different levels in the organisation. We conclude with Table 8 which provides a high level comparison of enterprise and project risk management.

Agile Risk Management

Abstract We argue that risk management in agile projects remains a passive and implicit activity that can be misdirected and often misunderstood. It is telling that whilst most developers have little difficulty explaining which features they are working (e.g., user stories) or to what level of quality they should be completed (e.g., definitions of done), few can comment on the capacity of their work to reduce (or exploit) project risk (Garland and Fairbanks 2010). We seek to address these shortcomings of agile methodologies by proposing a generic agile risk management process that embodies those aspects of traditional project risk management which lends itself to application in the spirit envisioned by the agile manifesto (Beck et al. 2001a). The objective of this process is to ensure that projects continue in spite of their risks. This is achieved by risk tailoring agile methodologies in the light of project specific circumstances and the wider enterprise risk management environment as well as identifying and treating the project risks on an ongoing basis. In doing so we attempt to overcome the natural and understandable reluctance of agilists to embrace something that might encumber their work practices. We underpin this process with research findings concerning the nature of IT project risk and provide a number of tools that can easily be embedded in an existing agile methodology. In treating risks we identify three principles (i.e., transparency, flow and balance) that capture the essence of the agile approach to risk management. Further, we assess existing agile techniques and evaluate their contribution towards risk management in order to provide advice on their effective deployment.

Agility and Risk

Although frequently mentioned, risk is treated in a rather narrow and implicit manner in agile projects. When we turn our attention to risk later in this chapter we will cite research identifying the main categories of risk in IT projects as project, schedule, supplier, people, requirements and technical. Yet consistently agile projects appear to

A. Moran, *Agile Risk Management*, SpringerBriefs in Computer Science,
DOI: 10.1007/978-3-319-05008-9_3, © The Author(s) 2014

focus almost exclusively on requirements and technical risk with few (e.g., DSDM) appearing to be aware of other sources of risk. Equally there is a prevailing perception that risk must inevitably be understood in terms of project threats (negative risk) and thus the opportunities (positive risk) presented in agile projects are inadequately managed. Yet nowhere more so than in IT does positive risk (in the form of opportunities) present itself through the value enabling nature of projects, the manner in which they are delivered and the operations and service delivery aspects that provide ample scope for exploiting synergies, stream lining processes and sustaining value creation (ISACA 2009a). Given the mantra of "embrace change" that pervades the agile community and in light of the enabling nature and value creation potential of IT projects, it seems ironic that such (positive) risk evaluation does not feature more strongly in agile projects (e.g., "embrace risk"). Fortunately, wider appreciation of risk is not entirely forgotten and is sometimes used as one of several factors used in the prioritization of task (Cohn 2010). Moreover process risk is a topic discussed at length in (Boehm and Turner 2009) though a more condensed version can also be found in (Boehm and Turner 2003). On the whole though, the agile community does appear to be lagging behind the risk management community, which for some time has understand its function as informing decision making in relation to the balance between risk and reward.

Perhaps the most commonly cited deficiencies include the lack of an explicit definition of risk, the focus on development aspects of risk (i.e., requirements gathering and technical implementation) without consideration of risks elsewhere in the SDLC, the nature of responsibility for risk, the recording and monitoring of risk, the environment in which the project takes place together with organizational attitudes to risk (Nelson et al. 2008) (which derives originally from work in (Nyfjord 2008)). Indeed many of the notions that ordinarily arise in the context of project risk management such as those described in the PMI Body of Knowledge (Project Management Institute 2013) and its more risk focused publications (Project Management Institute 2009), the Management of Risk framework (Office of Government Commerce 2007) which relates more to PRINCE2®and other sources (Hillson 2009), are simply nowhere to be found in agile project management. Given the growth in recent years of enterprise risk management, for example (ISO 2009), itself derived from (AS/NZS 2004) and (Moeller 2011), together with their application to IT (ISACA 2009a) this attitude to risk management seems parochial.

One of the most influential references to the embedding of risk management in iterative software development lifecycle was (Boehm 1986), further elaborated in (Boehm 1988), wherein a process model was described that actively seeks to identify and resolve risks and use these to influence the evolution of a solution from requirements through to operations. Though inherently iterative in nature, Boehm argued that his *Spiral Model* was not limited to a specific type of software development process (e.g., Waterfall, agile). Crucially Boehm argued that risk determines both the level of effort (e.g., perform product testing only to the extent that it reduces risk to an acceptable level) and the degree of detail (e.g., apply more design effort to innovative elements of a project) that risk related practices employ. These ideas were developed further into the notion of risk-driven architecture which advocates that the selection and application of techniques be motivated by risk and be commensurate with the risk of failure or success (Garland and Fairbanks 2010).

Whilst risk is frequently cited in XP (Wells 2009a; Jeffries 2013; Beck and Andres 2004) and some specific activities are noted in Scrum, neither formally address the need for and separation of identification, analysis and management activities. Studies of XP practices and their handling of risk are often critical of the confusion that can arise during release planning as developers struggle to negotiate value and priorities with customers (Li et al. 2006). This, it is claimed, leads to poor decision making, difficulties in balancing development risks and productivity and challenges in assessing risks. However, few offer concrete advice or guidance on tackling risk. For example, one proposal, (Li et al. 2006) to embed risk into XP is based on the notion of project profiles that capture such attributes as cost, schedule, quality and use these to determine user stories that can be combined in multiple feasible release plans based on their value, dependency, cost and required effort. The authors recommend the use of the Analytic Hierarchy Process, as expounded in (Saaty 1994), to automate this phase of planning, though their description of its applications appears to focus more on value maximization than on risk minimization. Risk management is then conducted by assessing the impact of risk on each release plan, though the authors restrict their understanding of risk to "losses caused by uncertain things ... [which] come from requirements, estimation or technologies" (Li et al. 2006, p. 424). Should no suitable release plan be identified then the process amends the project profiles and starts anew. It is suggested that this approach can be extended to longer term horizons provided that the project profile is reconsidered at the start of each iteration based on the outcomes of the previous iteration. This points to a diffusion of the risk management process over the iteration and increment cycles when viewed from the perspective of the generic agile software development process (see Fig. 4).

DSDM, which dedicates considerable attention to risk management, also frames it in terms of that "which may happen, and if it does, it will have a detrimental effect" (DSDM 2010). This notwithstanding, there is in DSDM at least an awareness of risk management concepts (e.g., use of a risk profile, provision of risk log templates) and some advice is given as to when and where risk related activities are appropriate. Moreover various whitepapers cite risk and provide guidance on how to tackle it (DSDM 2003c, b) much of which is implicitly or explicitly embedded in a wider risk aware environment (DSDM Consortium 2012a). Together these give an impression of a methodology already largely equipped with the mindset, if not the tools, to engage in risk management.

Generally speaking, agile methodologies make the claim that feedback mechanisms reduce (negative) risk by making information available in a timely manner, in order that the project members can advert negative outcomes. For example the formation of small heterogeneous teams, agile modelling and prototyping and the iterative and incremental nature of project work contribute in this fashion. Whilst these are indeed effective measures to reduce uncertainty, increase transparency and improve communication, there is little guidance on how these can be most effectively deployed. For example conscious prioritization of risks and their measures could improve the effectiveness of applied techniques, whilst clear identification of risks could lead to their clustering in such a manner as to improve efficiency of risk management. We will later make concrete suggestions in these regards when we introduce the agile risk management process and show how they might be applied in the Chapter "Applying Agile Risk Management". Needless to say these are merely guidelines and commentaries—ultimately it is the specific circumstances of a project and the inclination of its members that decide.

One aspect of risk that does receive some attention in agility is the balance of risk and delivery of value to customers when prioritizing tasks especially when doing so also takes into account other factors (e.g., cost). For example (Cohn 2010) advocates the strategy of addressing high value and high risk, high value and low risk and then low value and low risk tasks in that order, whilst entirely avoiding low value and high risk tasks. The argument is that working on high value and high risk tasks first eliminates significant risk early on. This approach treats risk as a facet of a task which might prove to be too limiting. For example some of the risks which a project must contend with are not inherent in the execution of specific tasks, but rather in the circumstances surrounding that execution and might otherwise be considered part of a project governance profile (i.e., the effective and efficient deployment of resources towards the achievement of the goals of the enterprise).

Why Risk Management Matters

Recall from the previous that project risk may be defined as uncertainty that has an impact on project objectives (Hillson 2009). In essence this implies that what is of interest is that which is unknown and which may impinge in a positive or a negative manner on one or more project objectives. As we shall see later, research suggests several prominent risk drivers of IT projects few of which are explicitly referred to in agile methodologies. Furthermore little to no explicit guidance is provided by most agile methodologies concerning the identification, recording or management of risks and there is practically no recognition of the cultural attitudes towards uncertainty or the management of risk at the enterprise level. This suggests that agile software development processes at best implicitly tackle risks and that those methodologies that lack a risk management framework suffer from the following deficiencies:

- *Inability to make informed risk and reward decisions*. A central function of risk management is the recognition of threats and opportunities within a project and to balance the desire for reward against the risks incurred in its pursuit. Accordingly an understanding of the risk appetite of a project together with the nature of the risks encountered in a project is central to such decision making.
- *Failure to identify appropriate risk response strategies based on risk exposure*. Risk exposure is a key determinant in the classification (and where appropriate prioritization) of risks. The inability to recognize risk exposure may therefore

impede the selection of an appropriate response. This point also encompasses the selection of agile techniques appropriate to the level of risk being managed.

- *Lack of oversight in risk monitoring.* Failure to engage in the monitoring of risk results in an inability to judge whether or not risk are being adequately managed. Team members ought to know how their activities are affecting project risk and how effectively and efficiently they are addressing risk. This shortcoming extends to enterprise risk management through the failure to identify projects that have overstepped risk boundaries in relation to the overall level of risk that an enterprise is willing to accept or is capable of absorbing.
- *Poor understanding of when to engage in risk activities.* Lack of understanding or inconsistencies about the perception of risk means that the responses to risk events will vary amongst team members who fail to explicitly agree on appropriate controls and triggers. Elements such as risk compensation (Adams 1995) and other cultural influences also come into play at this juncture.

Alternative models for integrating agility and risk management can already be found in the literature though these should be considered to be in their infancy (Nyfjord 2008)[1] or are specific to a particular methodology (Li et al. 2006). Studies such as (Nyfjord 2008) and subsequent research (Jaana Nyfjord and Kajko-Mattsson 2007, 2008), however, do warrant further attention owing to their comprehensive nature. First attention is drawn to the fact that agile methodologies fail to make specific suggestions for managing risk and that their capabilities require extension to cope with larger projects. This latter point is addressed particularly at XP and Scrum. It would therefore be interesting to validate these remarks against the claims made by more project oriented methodologies or the notion of scaling up (e.g., Scrum of Scrums, Disciplined Agile Delivery (Lines and Ambler 2012) or Scaled Agile Frameworks (Leffingwell 2007; SAFe 2013)) a topic that was already attracting attention in the years following publication of the manifesto (Eckstein 2004). In particular, the work outlined in (Jaana Nyfjord and Kajko-Mattsson 2008) advocates integration of agile and risk disciplines albeit in a qualified manner. The synthesis of traditional risk management with agile practices outlined in the body of that work highlights the tensions that arise when light and heavy techniques are mixed and may explain the adversity of the agile community reaction to such suggestions.

Agile Risk Management Process

We introduce the agile risk management process (see Fig. 11) as a means to risk tailor existing agile methodologies and practices in a manner consistent with the original spirit of the agile manifesto and in line with the risk context in which a specific project finds itself. Thus the process draws on elements of project and enterprise risk management and applies them appropriately to the agile project management.

---

[1] Which notes that it "does not describe *how* to conduct risk management in agile projects" and "does not provide guidelines for agile risk management" (Nyfjord 2008, p. 18).

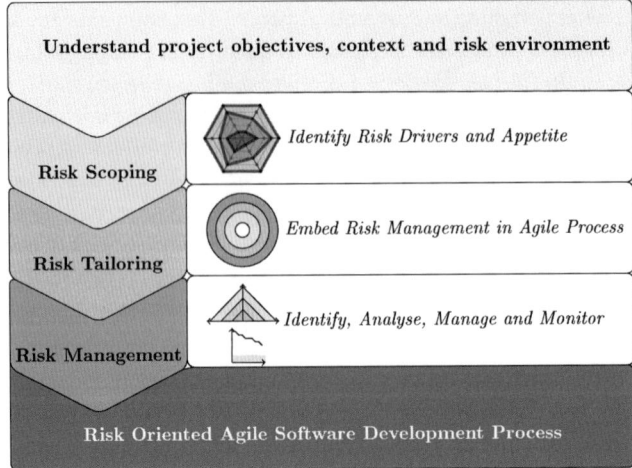

Fig. 11 Agile risk management process overview. Published with kind permission of © Alan Moran 2013. All Rights Reserved

The process begins with the definition of project objectives and context and maps these in terms of organisational risk management. This enables the project as a whole to be risk assessed alongside other similar projects in order to determine if its proposed benefits warrant the level of risk. Next, identification of risk drivers and determination of appetite is addressed in the pre-project phase though it may be returned to later if project circumstances alter significantly. Thereafter it is important to consider how the agile software development process being employed needs to be tailored in light of the risk environment of the project. This is a project (rather than methodology) specific undertaking since tailoring is an activity that is highly dependent on project parameters and circumstances. Accordingly we limit ourselves to high level guidance and advice. The remainder of the process is a reflection of existing project risk management practices adapted to the needs of agile practitioners.

Fundamentally the agile response to risk argues that tackling it involves deciding *what* should be done (i.e., a task, the completion of which reduces risk) or deciding *how* something should be done (e.g., deployment of an agile technique in a specific fashion). The former is a matter for the backlog (for which we suggest the use of *risk tasks*) and the latter is an adaptation of the process (for which we recommend agile charting and propose *risk tagging* as described later on in this chapter).

Roles

The suggestion that risk is a responsibility carried by all project members endears itself to the communal spirit of co-ownership and good citizenship that is expected of agile teams. This notwithstanding, the role of *risk manager* is sufficiently important for it to be assigned to an individual in the interests of ensuring that there is process compliance and that there is accountability for the effectiveness and efficiency of risk management activities. Thus it would seem natural that such a role be identified with that of the project manager (or loose equivalent) though the precise assignment of this role is less important than the fact that it is assigned at all. Suffice it to say that the existing roles of a given methodology ought to be augmented with that of risk manager. Thereafter we advocate that ownership of specific risk activities be based on existing pull mechanisms already used in agility for feature tasks (i.e., Kanban boards) or be taken up as auxiliary project activities undertaken by the project manager. This is a mild departure from traditional project risk management wherein ownership is assigned within a risk list early on in the process.

Principles

Throughout we shall be making reference to a number of principles that guide the agile risk management process which we refer to here for convenience. These principles work in concert with the values, such as openness, respect and courage, inherent in most agile methodologies and reflect features like communication and collaboration found in the principles of the agile manifesto.

1. *Transparency*. All risk related activities and artefacts should be visible to everyone in the team at all times. We propose that these be placed next to other agile tracking and reporting tools (e.g., burndown charts, Kanban board) or in a central project area (e.g., information radiator) and that team members be permitted to add to or annotate them. We refer to this practice as *risk walling* and it means that someone outside of the project could walk into the project area and immediately assess the risk situation without having to ask or interrupt team members.
2. *Balance*. Risk management is all about balancing risk and reward and finding ways of generating the same level of value with a lower level of risk. It should therefore be obvious which user stories bear the most risk and how the work of individual team members contributes to risk mitigation by either reducing threats or exploiting opportunities.
3. *Flow*. Risks are unavoidable in IT projects but understanding them and knowing how to deal with them enables the project to continue without serious disruption. For example, contingency plans agreed in advance make sure that should accepted risks materialize, the team knows what to do and is not interrupted with replanning or crisis activities.

Project Context and Risk Environment

We begin our discussion of the agile risk management process with understanding the project objectives, context and risk environment (see Fig. 11 to which we encourage the reader to return to throughout the discussion of the process). This stage takes

Table 9 Process: risk scoping

Input(s)	Activity	Output(s)
Project objectives and context	Gain an understanding of what constitutes risk in the project	Understanding of the nature of project risk. Amendments to project scope or definition of exceptions
IT risk drivers	Determine sources of risk and relate them to the enterprise risk environment	Validation (or initial creation) of enterprise risk driver map

place in the enterprise risk management context of the organisation and involves establishing an understanding of what the project is endeavouring to achieve and how much risk the enterprise is willing to tolerate in pursuit of these goals (see Table 9). This occurs during project initiation and requires that the project objectives be known which in turn provide the basis of what constitutes risk. We return to this matter later when we discuss risk identification so for now it suffices that the objectives are clearly stated.

Next the context of the project must be established since it is this that might exempt the project from ordinary controls (e.g., allowing more risk to highly innovative product launches) or otherwise constrain it (e.g., redefinition of project scope if deemed to be too risky). At this point in the process this must merely be noted and any decisions relating to the risk nature of the project be consciously taken and approval, where appropriate, be sought. The point here is not to invent bureaucracy but rather to be aware of projects that are being proposed which might be at odds with the enterprise risk (e.g., it is not appropriate in most industries to permit projects to engage in highly risky undertakings that could damage the reputation or financial standing of the organisation).

Finally we suggest that an understanding of the basic IT project risk drivers is required in order to properly assess a project proposal. Ideally this should be done at the enterprise level or failing that at the divisional, business unit or even programme level. Numerous research studies have identified various generic categories of IT project risks (a good survey of these can be found in (Arnuphaptrairong 2011) which also includes the widely cited (Boehm 1991)). Table 10 borrows categories from (Ropponen and Lyytinen 2000) but reorganizes them and integrates much of the findings from the other studies to produce a list of generic IT risk drivers that may help as a starting point.

When grading risk drivers, a simple scale comprising of five to seven points is often adequate. The grades should be expressible in business terms and be understood by all participants. Table 11 is a simple five point scale that describes the range of risks that might be encountered in the technical risk category. Always bear in mind that risk can have upside as well as downside. Thus technological uncertainty could be considered as much an opportunity as a threat and this may be reflected in the willingness to embrace risks if there is a perception that this may bring advantages. Therefore care should be taken not to frame the grading system in unduly negative

Table 10 Archetypal risk drivers for IT projects

Risk driver	Description
Requirements risk	All risks relating to functional requirements and the changing nature of requirements in general. Issues of user acceptance might also be relevant to this category depending on the nature of the project. For the purposes of this thesis we have subsumed business risk into this category. Remaining non-functional requirements not already accounted for under technical risk may be included here (e.g., usability, security)
Technical risk	All risks relating to architecture, design and infrastructure of the proposed solution. This risk is not limited to the developed solution but also encompasses dependencies (e.g., shared libraries) together with the estimation of hardware and software dimensioning and capabilities. Typically the majority of non-functional requirements are included here (e.g., maintainability, scalability, performance)
Schedule risk	All risks arising from the scheduling and timing of activities (including the release planning of increments) and the financial cost consequences thereof (e.g., net cash flows, investment decision making)
Project risk	All risks relating to the effectiveness of the project management methodology, levelling of resources and the management of project complexity. Depending on circumstances it may also be appropriate to include risks relating to management support for the project
Supplier risk	All risks relating to external sourcing including consulting and delivery of components (incl. timeliness, conformity, quality). It is not uncommon to subsume risks relating to the stability, continuity and capability of suppliers into this category
People risk	All risks related to the level of skills in the team and expectations of abilities. These risks are often impacted adversely by other risks as impending project deadlines place greater stresses on staff

terms. Insofar as this is possible a standardized set of drivers and scales ought to be used throughout the enterprise and embedded in this should be the consciousness of enterprise risk management environment.

Once a graded list of risk drivers is available it is time to reach a consensus concerning the upper and lower risk thresholds for each of the risk drivers. The *upper risk threshold* is that level beyond which urgent and immediate risk mitigation activity would need to be initiated. This embodies an uncomfortable level of uncertainty with which the organisation would have difficulty coping with. The *lower risk threshold* is that point below which the risk is considered negligible and scarcely needs further monitoring. This is the comfort zone and the realm of daily business. The region between the two thresholds represents risk that ought to be monitored and acted upon at an appropriate time if deemed necessary. For example in Table 11 an enterprise might feel that the levels "Terra Firma" and "Cautious Explorers" hardly constitute concern though might be conservatively minded enough to consider the risk surrounding "Early Adopters" to be a step too far.

An *enterprise risk driver map* is created by connecting together all the upper respectively lower risk threshold points and marking the region above the upper limits red, the region below the lower limits green and the remaining region yellow (Fig. 12).

Table 11 Sample five point (descending) scaling for technical risk

Scale	Description
Market makers	Highly innovative and ground breaking technology that requires new and perhaps unproven technologies or application of existing technologies in unanticipated fields
Early adopters	Inroads into technologies that have seen some industry use though a significant effort must be expended to become capable and effective in their use (e.g., tool support might be immature)
Forward movers	Significant technical innovations on several fronts where industry best practices and reference implementations already exist. Technical practices may require interpretation and adaptation to the organisation
Cautious explorers	Mainstream technical implementations that require some evaluation of new technologies integrated into existing frameworks and platforms. Infrastructure is established but training in its use might be required
Terra firma	Well tried and trusted technologies in which the organisation is highly invested. Tools and processes are well known and adhered to

Thus a risk driver map is a visual indicator of the attitude towards risk at the enterprise level. The value of a risk driver map becomes clearer when the personal attitudes towards risk of the project team members are taken into account (see the Chapter "Project Risk Management"). Willingness to take more risks or the adoption of a risk adverse stance may prove to be at odds with the project risk appetite and thus it becomes necessary to clarify what constitutes acceptable and unacceptable risk. It is at this stage in the process that the project manager should take the time to understand the personal risk attitudes of team members.[2]

As we shall presently see, by plotting individual projects on the risk map, we can use it as the basis for assessing whether or not to proceed with them or to decide if exceptions need to be considered. The value of the consistency of an enterprise risk driver map is that it enables knowledge gained in assessing risk appetite from completed projects to be transferred over to new ones. Indeed if an enterprise risk management framework (e.g., ISO 31000 (ISO 2009), COSO (Moeller 2011) or Risk IT (ISACA 2009a)) is already in place, then there should be some form of linkage between project risk appetite and the wider corporate attitude towards risk.

Risk Scoping

Using the enterprise risk driver map it is necessary to assess the how the drivers affect a specific project. Such discussions ought to involve the project sponsor, team members and any additional primary stakeholders, who together attempt to gain a

---

[2] For example, DSDM promotes the use of a Project Approach Questionnaire that could be adapted for this purpose.

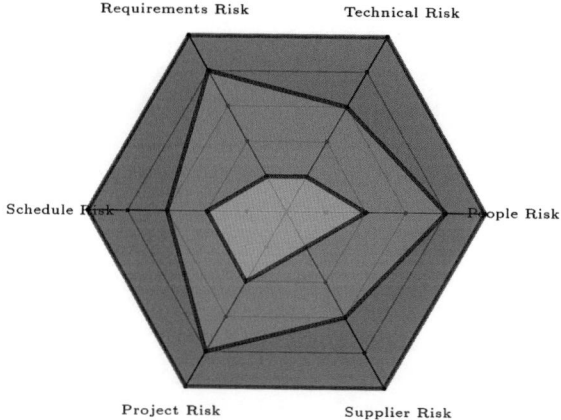

Fig. 12 Risk driver map based on archetypal risk drivers. Published with kind permission of © Alan Moran 2013. All Rights Reserved

Table 12 Process: risk scoping

Input(s)	Activity	Output(s)
Project context and enterprise risk driver map	Determine those risk drivers that are most relevant to the project	Project risk driver map

common understanding of risk. A facilitated workshop would be the appropriate forum for such a discussion (Table 12).

Figure 13, which shows a *project risk driver map*, illustrates a possible outcome of such a workshop from which we can see that the technical demands of this project are dangerously high for the organisation but supplier risks are rather low. This might be the profile of a project in a company that has up until now been focused on developing web applications but is faced with a project to create a mobile app. For this company, technical risks pose a challenge since this is a new field for them. However, established suppliers are available since mobile apps are not a new technology suggesting that it might be possible to translate technical risk into (lower) supplier risk. Alternatively, a longer term strategy to reduce such technical risk would be to build up mobile app development skills in-house. Thus the idea is to encourage a serious assessment of who is doing the project and whether preconditions should be met before the project is approved (e.g., finding suppliers). In this example the remaining risk drivers require monitoring but are not grounds for stopping the project.

Should the project risk driver map indicate that the project is on the enterprise boundaries of intolerable risk then this will have consequences for how the project is conducted. It may become necessary to choose appropriate tools consistent with risk-driven approaches as described in (Garland and Fairbanks 2010) which advocate that the selection and application of techniques be motivated by risk and be commensurate

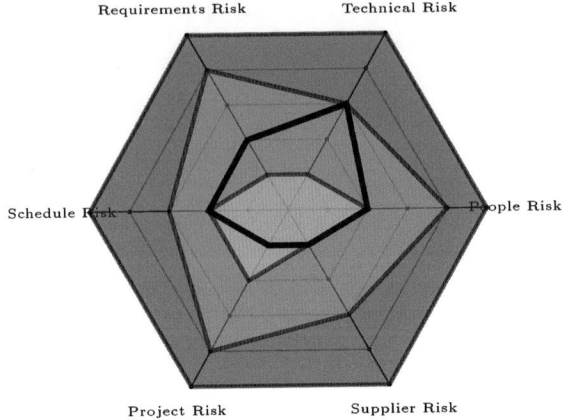

Fig. 13 Project risk scoping based on enterprise risk driver map. Published with kind permission of © Alan Moran 2013. All Rights Reserved

Table 13 Process: risk tailoring

Input(s)	Activity	Output(s)
Choice of agile methodology	Create an agile chart for the chosen agile methodology	Agile chart
Agile chart and project risk driver map	Embed risk management activities into the agile chart based on the project risk appetite	Risk adjusted agile chart

with the risk of failure (or success). For example, if developing an application that must deliver a high performance using a new algorithm then it might be necessary to invest more heavily in test driven development throughout the project (e.g., perhaps including some performance tests into continuous integration).

Risk Tailoring

The next stage of the agile risk management process concerns project specific amendments to the agile methodology which can only be done in light of an understanding of the risks facing the project already identified in the risk scoping stage. Up to this point in the process we have not explicitly considered which agile methodology we are using, however, it is now necessary to begin considering the specifics of our chosen methodology. For now, our remarks remain relatively general in nature though we will return to this topic in more detail in the Chapter "Applying Agile Risk Management" (Table 13).

Having an agile chart (see the "Agile Software Development" chapter) for the software development process is in itself a useful communication tool that fosters a better understanding of the cyclical nature of activities. It helps clarify when and how often specific activities are expected to take place. Indeed this has an impact on the manner in which risk management is performed (e.g., does the majority of planning occur at the incremental or iterative level, when should risk assessment activities take place?). Some practitioners like to annotate their agile charts with the artefacts of their chosen methodology creates (e.g., product vision, iteration backlog) in which case it would also make sense to add the risk artefacts (e.g., project risk driver map). When considering where to place risk management activities it is important to consider the state of available information. For example, it may be tempting to conduct a risk analysis at the increment level but if there is insufficient information to evaluate risks properly then it is more appropriate to perform this analysis at the iteration level. This is a matter of judgement that is best left to the team to decide for itself— but the decision should be explicit and made clear at this stage in the process. Agile techniques themselves (e.g., prototyping, continuous integration, refactoring) should also be considered risk management tools though how they are deployed depends on the necessary level of information required together with the desired frequency and intensity of the activity.

Figure 14 is indicative of what risk tailoring of the generic agile chart presented in the "Agile Software Development" chapter might look like (see Fig. 4 though we shall explain the tools mentioned here later in this chapter). This is intended as an illustration of how to risk tailoring should be understood in general terms. We return to specific applications of common methodologies (i.e., XP, Scrum and DSDM) in the Chapter "Applying Agile Risk Management".

Risk Management

The next stage of the agile risk model concerns the identification, evaluation and management of risks at the project level. For the most part this entails the application of traditional risk management techniques though some agile adaptations nonetheless apply. The main purpose of risk management is to heighten the understanding of risk and return and to support decision making in this respect within the project context (Table 14).

Conscious of the project parameters such as time and budget, most project managers are understandably worried about deviations that take them outside of project tolerances and might instinctively understand risk in negative terms. Agile project managers, however, reject the full up-front plan-driven paradigm and should perceive uncertainty both as an opportunity and as a threat. As already commented, there is nonetheless a tendency in agile literature to focus entirely on the negative aspects of risk particularly in relation to requirements and technical risk. In this section we encourage a wider perspective on risk management that addresses both risk taking and risk avoidance stances. Throughout the risk management process it is

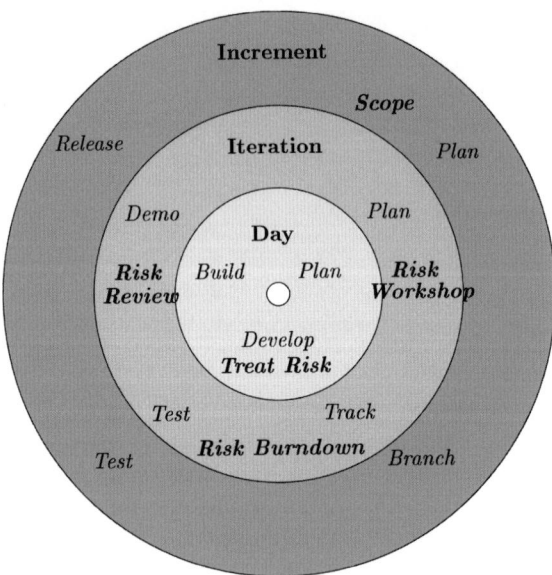

Fig. 14 Risk adjusted generic agile process (compare with Fig. 4). Published with kind permission of © Alan Moran 2013. All Rights Reserved

Table 14 Process: risk management

Input(s)	Activity	Output(s)
Project risk driver map	Identification of risks commensurate with the available level of information	Risk list
Risk list and risk pyramid	Analysis and prioritization of risks in terms of likelihood and impact	Risk list
Risk list and backlog	Management risks in terms of response strategies and tasks	Risk list, updated backlog (incl. risk tagging) and risk modified Kanban board
Backlog and risk list	Monitor and track risk related activities. Observe possible changes in underlying risk exposure components	Risk Burndown (see later)

recommended to maintain a *risk list*[3] that underpins all risk based activities and should include the attributes listed in Table 15. Note that the risk owner is deliberately absent from the register for two reasons. The first is that when the measure is a task then it must be treated like any other task which means that it is placed on the backlog and any project team member may assign themselves the task. Secondly not all risk measures can be assigned an individual owner (e.g., the risk measure to

---

[3] We consciously avoid the term *risk register* to hint at the lightness of our means of recording risks.

Table 15 Common risk list attributes adapted for the agile context

Attribute	Description
Identifier (optional)	Code or name identifier with which the risk may be referred to. Unless this is actually used somewhere it ought to be considered optional
Description	Description of the precise nature of the risk as elicited in the risk identification phase. The use of a *risk statement* (see below) is highly encouraged
Classification (optional)	Placement of the risk in some form of categorization. Typically this is based on risk drivers though the project may choose to derive a classification based on other factors. Once again this should only be used if considered useful (e.g., tracking based on category), otherwise it should be dropped
Likelihood	The probability with which the risk event is likely to occur or its proxy, frequency (with respect to a specific timeframe). This information can only be gathered during the risk analysis stage
Impact	The impact the risk event, should it occur, is likely to have. This information can only be gathered during the risk analysis stage
Score	The risk score assigned on the basis on likelihood and impact (see below)
Strategy	The risk strategy to be adopted in order to manage the risk. This information can only be gathered during the risk management stage
Measure	The measures (if any) to be taken to address the risk. Note that a reference to a measure can be a task (e.g., introduce redundancy into the storage) in which case a risk task as described later must be included on the backlog or it may concern the way something is done (e.g., all GUI related tasks must use pair programming) in which case it ought to be flagged up in a manner that all are reminded of it (e.g., risk tagging)
Residual likelihood	The resultant likelihood of the risk event once the measure has been implemented
Residual impact	The resultant impact of the risk event once the measure has been implemented
Residual score	The risk score assigned on the basis on residual likelihood and impact (see below)

employ a specific agile technique) in which case it is the responsibility of all who are affected by the decision.

We suggest that the risk list be made visible to the team at all times (e.g., placed next to the burndown charts) in order to permit anyone at any point in the project to add new risks as they come to light. Periodically the risk list will need to be reviewed, risks analysed and consolidated. Risk management is a continual process that requires revalidation of the premise on which it is based (e.g., risk appetite). It is recommended that risks be identified at the start of each iteration and monitored throughout in all but very low risk projects where risk management at the incremental level might suffice. Prioritization is influenced by risk strategy. For example some risk measures are proactive in nature i.e., an undertaking is committed to in order that an inherent risk is reduced to a residual risk and such measures can be prioritized in the normal fashion. Other measures (arising from the accept strategy) are reactive in nature and are only engaged in once the risk has been realised. Thus we must accept that their priority is contingent on the risk being triggered. Here we need to be at least aware of the *contingent priority* of the risk action i.e., how urgent must the action be undertaken were the risk to be realised and then plan accordingly.

One comment relates to non-task risk activities (e.g., application of an agile technique to reduce risk) and affects how estimation should be performed. Such a decision impacts tasks on the backlog and thus triggers a reassessment of their estimates which must absorb the consequences of the decision to undertake tasks in a specificfashion. For example, suppose that the risk of implementing a GUI suggests that pair programming is a necessary risk mitigation measure, then this decision (which in effect adapts the agile process) might entail overhead[4] that would otherwise not have occurred had the decision to deploy pair programming not been taken. Thus one must apportion this overhead to each of the tasks based on their sizes.

Risk Identification

Risk identification is the activity of determining the main uncertainties that could have a plausible and material effect (positive or negative) on the project objectives. Note that it is not necessary to identify every conceivable source of uncertainty. The project risk drivers constitute a good starting point for risk identification though other risks may also be found. This requires experience and judgement as well as attention to a couple of the subtleties of risk identification.

Firstly, it should be noted that whilst there may be many uncertainties surrounding a project, not all are in fact relevant. To repeat an earlier metaphor the outcome of a horse race is always uncertain, but only becomes a risk once a bet has been placed on a horse. Thus uncertainties that do not impinge on project objectives can simply be ignored.

Secondly, confusion often arises when attempting to distinguish between risks and non-risks, in particular causes and effects. A project task is driven by its rationale (cause) and is subject to uncertainty (risk) that may influence its outcome (effect). The confusion arises mostly commonly between risk and effect and is best illustrated using an example. Consider the migration of a website from one server to another. It is plausible to cite the lack of availability of the website following the migration as a potential risk. This, however, is not a risk but rather an effect i.e., either the website is available or it is not. The real source of the risk can be found by identifying where the uncertainty lies (e.g., lack of knowledge about how DNS entries should be reconfigured). What is so pernicious about this misunderstanding is that it often directs risk management activities towards actions that fail to address the underlying risks. In this example were non availability considered the risk then this might motivate activities such as the deployment of failover hardware. The correct

---

[4] We say *might* since some studies, (Williams et al. 2000), have shown that pair programmers are more efficient than individuals working alone i.e., the sum is truly more than its parts!

response, however, would be to tackle the uncertainty inherent in the migration (e.g., train an administrator in DNS configuration).

The following are two common techniques that can be easily applied in the agile context to identify risks:

- *What/Why* In this technique the question is asked *what* could happen. It is important to resist the temptation to always frame this question negatively (e.g., "what could go *wrong*?") but rather to admit a neutrality and openness to the notion that there are positive as well as negative risks. Contained in the follow-up *why* question, are the risks that really need to be addressed. This approach is particularly useful for brainstorming workshops especially where participants have relatively little experience with risk identification.[5] Applied to the website migration example earlier this might result in an answer to the *what* question of the form "the website in no longer available" which could solicit the follow-up explanation "we're not sure how the DNS entries need to be updated". Central to this technique is to watch out for wording that suggests uncertainty (e.g., "not quite sure about...", "uncertain about...").
- *Risk Statements* This approach advocates the formulation of structured statements of the form "*As a result of <definite cause>, <uncertain event>[6] may occur, which would lead to <effect on objective(s)>*" (Hillson 2009). This approach can be very effective though it does require some practice to master completely. Thus the website example above might produce a statement such as "as a result of the need to migrate the server, a DNS misconfiguration might occur, which would lead to the website no longer being accessible".

From the outset it should be clear that we are only tackling first order risks where cause and effect can be directly linked. Reality is, of course, more complex and the link between cause and effect may traverse a series of interdependent nodes each representing intermediate risks that must be addressed. It is, however, beyond the scope of this thesis to consider this matter in detail and thus we limit ourselves to illustrating the point that it is risks not effects that we are interested in.

It is not uncommon to see both techniques above being used together. We suggest starting with a "paper round" approach where the risk identification workshop is divided into two stages (see Fig. 15). During the first brainstorming stage "whats" are written as titles on separate pages and these are then reviewed to eliminate duplicates and clarify any misunderstandings. Then during the second phase, these pages are handed around in a circle to each team member who is free to write down any potential "whys". These pages can then be gathered up by the risk manager and used to derive the risk list entries using the risk statement technique. Combined these approaches

---

[5] An interesting variation on this approach is the DSDM recommendation of using a "project obituary" where at the start of the project, team members must imagine which events might have led to a fictitious failure of the project (DSDM 2010).

[6] Eventuality is a core concept in this formulation that draws attention to the notion that the circumstances of risk realisation must be testable i.e., a specified threshold must have been breached beyond which action is to be instigated.

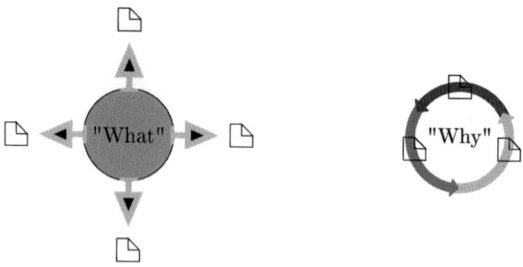

Fig. 15 Risk identification based on What/Why technique. Published with kind permission of © Alan Moran 2013. All Rights Reserved

Table 16 Example risk list with risk identification information

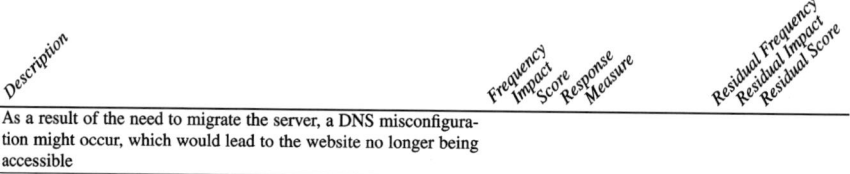

Description	Frequency	Impact	Score	Response	Measure	Residual Frequency	Residual Impact	Residual Score
As a result of the need to migrate the server, a DNS misconfiguration might occur, which would lead to the website no longer being accessible								

can be a very effective and efficient means of gathering input from a wide range of stakeholders. Table 16 illustrates an extract from the risk list that records the website migration risk.

Risk Analysis and Prioritization

Once risks have been identified it is necessary to evaluate and prioritize them. Risk assessment involves estimating both the *likelihood* and the *impact* of risks, the issues surrounding which have already been alluded to in the Chapter "Project Risk Management". Note that impact assessment should be in terms of project objectives and that these may already themselves have been prioritized (e.g., MoSCoW) in which case this should have a bearing on the risk assessment. Recall from the Chapter "Project Risk Management" that we presented the traditional representation of a risk map as found in (ISACA 2009b) and other sources (Fig. 16).

The disadvantage with this diagram is that it is based on absolute exposure and thus places opportunities and threats together in the same region. Instead we propose that opportunities and threats be displayed separately reflecting the fact that IT projects should record and be open to the notion of positive risk which should at least be brought to the attention of the project sponsor. Note that the vertical axis in Fig. 17 is likelihood and the horizontal axis is (positive or negative) impact which is the opposite to the traditional risk map representation.

Tracking risk requires that risk assessments be reduced to a comparable basis. Risk exposure would appear to provide a reasonable means of achieving this. However, the fact that risk components are not always quantifiable but may be ordinal in nature

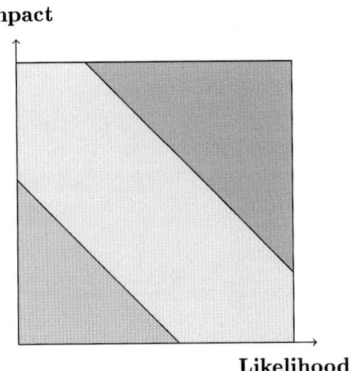

Fig. 16 Traditional risk map. Published with kind permission of © Alan Moran 2013. All Rights Reserved

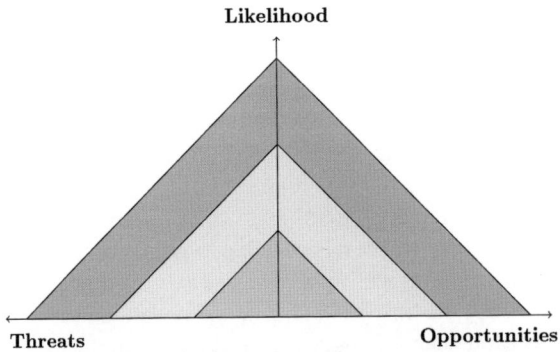

Fig. 17 Risk pyramid of threats and opportunities. Published with kind permission of © Alan Moran 2013. All Rights Reserved

means that exposure (i.e., the numerical product of components) may not make much sense. An alternative is to assign "risk points" analogous to the use of story points in estimation. This would imply that similar risks ought to have similar point values though what constitutes comparable in such situations may be far from clear. The proposed approach suggested here is to assign a score to each band found on the risk pyramid by according green (the innermost pyramid) risks a single point, yellow (the middle pyramid) risks two points and red (the outer pyramid) risks four points. As circumstances dictate this might be refined further (e.g., four points for the inner side of the red band and six points for the outer band). This incorporates the element of risk exposure whilst recognizing the multiplicative nature of its components. In order to avoid any potential misunderstandings the term "risk points" is eschewed in favour of *risk scores* underlying the crude nature in which they are derived.

Returning to our earlier example of a website migration we judge the likelihood of a misconfiguration to be relatively small but the impact large. This places the risk in the red region (bottom left corner of the risk pyramid) and accordingly the

Table 17 Example risk list with assessment and prioritization information

Description	Frequency	Impact	Score	Response	Measure	Residual Frequency	Residual Impact	Residual Score
As a result of the need to migrate the server, a DNS misconfiguration might occur, which would lead to the website no longer being accessible	S	L	6					

risk is scored 6. We return to the risk list to add the new analysis and prioritization information (see Table 17).

Prioritization of risks can be performed in a number of ways. For example, it is not uncommon to treat as important risks associated with high value items on the grounds that even a relatively low risk on a high value item might have higher priority that a high risk on a low value item. Others recommend a specific order of prioritization in which risk and value need to be balanced such as addressing high value and high risk first, then high value and low risk and finally low value and low risk tasks whilst avoiding entirely low value and high risk tasks (Cohn 2010). Occasionally risk and value are directly linked in a manner using the argument that "more often than not high risk work items also prove to be of high value" (Lines and Ambler 2012) though we prefer to suggest that high risk items must be of high value in order for them to merit consideration. Finally whilst it is possible to prioritize risks purely in terms of descending risk exposure irrespective of business value, we advise against this as to do so would be to break the link between risk and reward that is central to effective risk management. Moreover we suggest T-shirt sizing (i.e., XS, S, M, L, XL) be employed and encourage priority to be inferred from the colour of the region into which the risks land on the risk pyramid and to link these to business value. This is the essence of the balance of risk and reward and suggests that risk prioritization should never be an isolated assessment.

Risk Treatment

Risk treatment is concerned with the selection of response strategies to risk and determining the nature of measures to be taken to reduce the risk. This in turn informs the specific measures to be undertaken in the risk activity. It is important to make an estimate of the residual risk taking into account this measure in order to gauge the effectiveness of the proposed measure. Indeed if the residual risk is still considered too high (when the project risk appetite is taken into account) then either alternative or additional measures must be decided upon. There are essentially the three options available:

1. *Risk Tasking*. Risk tasks be formulated, linked to items on the risk list and placed on the iteration backlog.

2. *Risk Tailoring/Tagging*. The decision is taken to manage risk through the application of an agile technique to specific tasks or to integrate them into the overall project methodology (i.e., modify the project agile chart).
3. *Contingency Planning*. Determine what activities would need to be undertaken for a risk that has yet to materialize but which we are nonetheless willing to accept. These are contingency tasks which "lie in wait" on the backlog but which can be expected to rise in priority if the risk were to occur. For example, if we were to accept the risk that a data migration might not complete as expected we might have the contingency plan to recover lost data for which a certain amount of work must be undertaken.

Of the three options above, risk tasking is the most straightforward as it involves defining specific actions that need to be undertaken that are often directly linked to user stories (e.g., altering the design of a piece of code or ensure higher redundancy in infrastructure components). If using a Kanban board, then we recommend that risk tasks be coloured appropriately to make them visually distinct from other tasks. The second point above refers to the application of agile techniques specifically to tackle risk. For example, GUI work might be performed using the pair programming technique with the customer sitting together with the developer.[7] Thus this might apply to several user stories, though it suffices to record the technique on the risk list (rather than the affected stories or tasks). We suggest that individual tasks be *risk tagged* with the relevant techniques (e.g., by attaching coloured symbols to the affected tasks). Sometimes a risk technique will be deemed to have wide applicability and becomes part of the agile chart (e.g., the adoption of a test driven development culture). In such cases it suffices to embedded the practice into the project methodology making it unnecessary to record on the risk list. Finally there may be circumstances in which we accept a risk and must be prepared should it be realised. In such cases we create a *contingency plan* as a form of conditional task that is only performed if necessary. Priority escalation must be determined at this point too since it likely that there will be some urgency connected to its execution should the risk actually materialize.

Figure 18 illustrates a simple application of the above options to three user stories and their associated tasks (note that we omit certain details such as story points and risk scores for reasons of simplicity). Both user stories U1 and U2 have explicit risk tasks (T1.3 and T3.2 respectively) that address specific risks in those stories. Tasks T1.1, T2.2, T3.1 and T3.2 have all been tagged for pair programming (e.g., perhaps because the benefit of the immediate feedback is sufficiently risk mitigating) where tasks T2.2 and T3.3 are tagged for prototyping that requires customer review (e.g., to reduce requirements risk). In the risk list we link these tags to the underlying risks they address but for reasons of brevity do not cite every individual affected story. Finally user story U3 also includes a contingency, task T3.3, suggesting that there is risk present that we are willing to accept. Provided that this is not related to T3.2 (e.g., perhaps we are unsure about whether or not our existing risk measures will

---

[7] The term "non-solo" development is used in (Lines and Ambler 2012) to generalize the notion of "pair programming".

Fig. 18 Risk modified kanban
board. Published with kind
permission of © Alan Moran
2013. All Rights Reserved

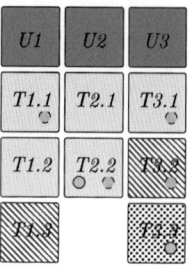

be effective) then this might be deemed acceptable. However, U3 does strike us as
rather risky and it would be worth asking whether or not the same value could be
delivered by a less risky route.

The practice of making all risk related artefacts visible on the project wall, which
we refer to as *risk walling*, is an expression of the both the transparency and the
balance principles and typically involves the project risk driver map, the risk list, the
risk modified Kanban board, the risk burndown chart (which we introduce later) and
the project agile chart. Together these communicate risk management attitudes and
activities in a highly accessible, understandable and visible manner.

When it comes to the selection of agile techniques there are essentially two options
open to the agile project manager. Either a technique must be applied with greater
intensity or with greater *frequency*. Given the fixed timeframes of iterations we can
really only speak of one degree of freedom that links these two approaches since
intensive activities require more time and can therefore be conducted less frequently.
Generally speaking when one considers this matter in terms of an agile chart then by
definition more frequency based responses are generally belong in the inner cycles
whereas intensity based responses are more often found in the outer cycles as depicted
in Fig. 19. By way of illustration consider the handling of product quality risk where
compliance to specific coding standards is the objective. An agile team may be
faced with either conducting code reviews at the end of each iteration (intensity)
or adopting pair programming throughout the iteration (frequency). Either way the
same goal is arrived at albeit through different routes—which is more effective is a
matter for the team itself to decide but this should be openly discussed and explicitly
decided upon. Should the project risk appetite suggest a high tolerance of risk then it
becomes appropriate to relax the intensity and frequency of certain agile techniques.
For example, if an implementation relies on standard design patterns requiring little
new innovation then it is appropriate to dispense with prototyping and attempt a direct
solution relying perhaps entirely on test driven development, continuous integration
and refactoring practices. In effect the process is relieving itself of some of the risk

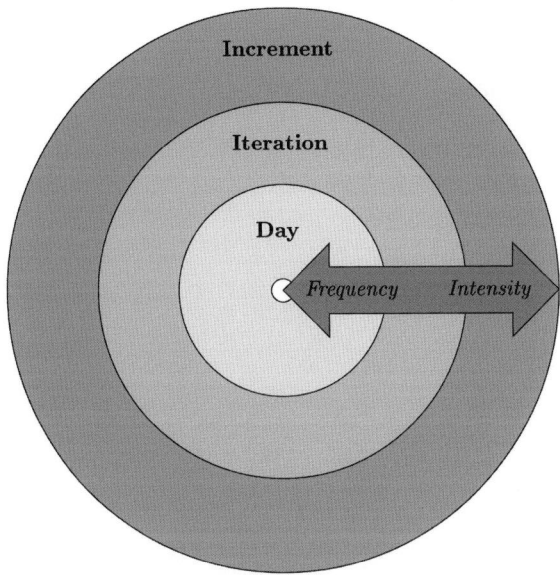

Fig. 19 Single degree of freedom in timeboxed environments. Published with kind permission of © Alan Moran 2013. All Rights Reserved

controls that would otherwise have been necessary but in this case would have been too costly or not warranted given the level of risk involved.

Figure 20 is a broad indication of where some of the agile techniques cited in Appendix A might be placed in relation to their frequency and intensity. It ought to be noted that the precise nature of these practices varies according to methodology and thus this diagram may require reinterpretation in specific circumstances. Some activities can afford to be repeated infrequently if they are automated (e.g., continuous integration occurs may times in the daily cycle without much involvement from programmers) and others are simply too hard to place precisely on the diagram (e.g., prototyping) since they can be practiced at varying levels of frequency and intensity (Fig. 19).

Given the most commonly cited risks in agile communities (i.e., requirements and technical risks), the usual technique based strategies including iteration planning with heterogeneous team (iteration cycle), customer demonstrations and retrospectives (iteration cycle), prototyping (iteration or day cycles), daily builds (day cycle), pair programming, continuous integration, test driven development and perhaps refactoring (day cycle) constitute a robust and varied toolset. That said, what might occur less to an agile team, is to use heavier techniques not ordinarily associated with their repertoire. Thus, for example, Business Process Modelling (BPM) is an excellent tool for gaining consensus on the operational details of a process and the use of quality function deployment (otherwise known as "House of Quality" owing to the graphical shape of its matrices) is a well-proven quality technique for mapping requirements

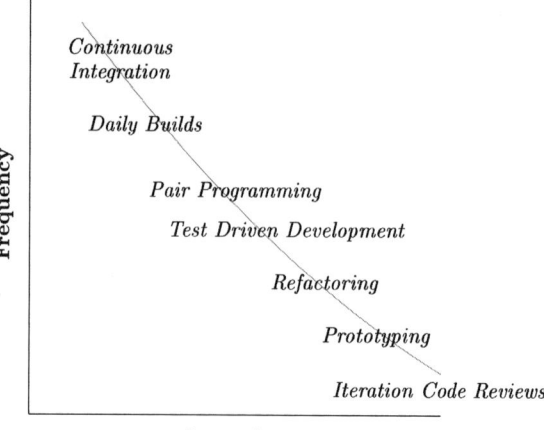

Frequency

Continuous
Integration

Daily Builds

Pair Programming

Test Driven Development

Refactoring

Prototyping

Iteration Code Reviews

Intensity

Table 18 Agile interpretation of the classical risk strategies

Risk strategy	Description
Accept	No specific action will be undertaken to mitigate or manage the risk. At most a contingency task could be defined and assigned a low priority (e.g., Could in the MoSCoW scheme). Unfortunately most agile methodologies are relatively silent on the topic contingent tasks though some suggest use of "collective" tasks (e.g., rather like bug fixing) whilst others allocate contingency through prioritization (e.g., MoSCoW)
Exploit	Measures will be undertaken to increase either the frequency or the impact of the event. The decision to act on the risk should be taken during iteration planning and a task should be assigned with an appropriate definition of done. Alternatively the manner in which an existing task is to be performed might require redefinition (e.g., use of a more flexible design pattern might be advisable if warranted by risk)
Share	Measures will be undertaken (with others) to share the risks and rewards. As with exploit explicit task planning should be undertaken though perhaps with lower priority since shared risks are often those that the project team alone cannot bear
Transfer	Undertakings to ensure that responsibility for the risk and its treatment are passed to a third party. For example, this could constitute some form of compensation (e.g., insurance) or outsourcing (e.g., use of external specialists)
Reduce	Actions will be undertaken to reduce either the impact or the frequency of occurrence of the event. As with exploit an explicit task should be defined and included in iteration planning or existing tasks augmented (e.g., production of prototypes, use of workshops to clarify requirements further)
Avoid	The activities that give rise to the risk will not be undertaken. Thus the offending tasks should be removed entirely from the iteration plan

to architectural facets. Such techniques, which fall into the intense category, merely provide a forum for discussion and ought to be pursued only in so far as they address issues of uncertainty.

The natural agile placement of the risk response strategies listed in Table 18 in the risk pyramid is depicted in Fig. 21. This ought to be the first port of call

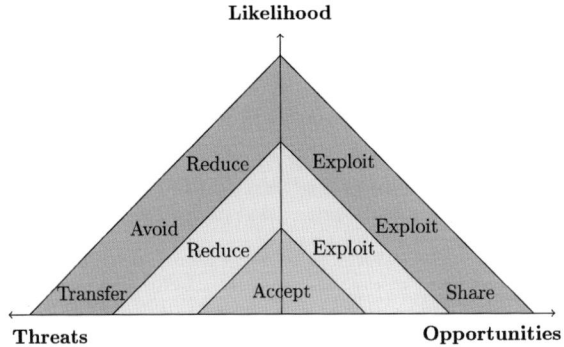

Fig. 21 Risk response strategies determined by risk exposure. Published with kind permission of © Alan Moran 2013. All Rights Reserved

when deciding how to deal with a risk when exposure is the primary determinant. Once a risk response strategy has been determined it is necessary identify one or more specific responses. It is important to understand that risk measures are an integral part of the activities during an iteration and should be treated no differently from other tasks (e.g., features, quality). Thus we strongly advise against the maintenance of a separate "risk backlog" as this makes no more sense than any other form of specialized backlog (e.g., quality backlog, maintenance backlog).

Returning to the website migration example we note that our risk pyramid suggests that we transfer the risk (e.g., by finding a system administrator outside of the project team to perform the necessary configuration for us). At this point it is worth considering how sensitive our risk response strategy is to the underlying assessment. For example, if we felt that the risk was medium rather than high then we would endeavour to reduce the risk (e.g., by adding a task to learn about DNS configuration) and if we felt that the likelihood was higher we might try to avoid the risk entirely (e.g., by not migrating the website). Let us assume that on reflection we stand by our original assessment and accordingly we update the risk list (see Table 19) where T1 refers to a task concerning the outsourcing of DNS configuration to an external expert. We are expressing the belief that this measure alone suffices to reduce the risk to an acceptable level (i.e., risk score of one). Note that the risk is not entirely eliminated since although we may be confident that the technical risk is negligible we have introduce other forms of risk (e.g., supplier risk) which must be accounted for.

In summary a risk score reflects the current level of risk, a strategy suggests the general direction in which action should be taken, a measure is an undertaking to reduce the score and the residual score is the anticipated level of risk were the measure to be enacted.

Table 19 Example risk list with measure and residual risk assessment

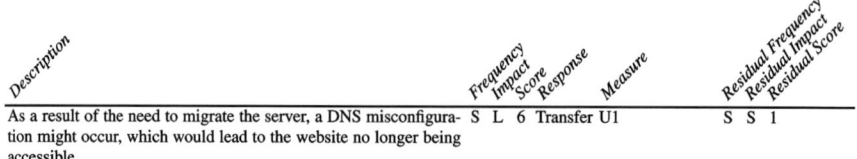

Description				Frequency	Impact	Score	Response	Measure		Residual Frequency	Residual Impact	Residual Score
As a result of the need to migrate the server, a DNS misconfiguration might occur, which would lead to the website no longer being accessible				S	L	6	Transfer	U1		S	S	1

Risk Monitoring

The most effective means of monitoring risk reduction is through the use of a *risk burndown chart*, see Fig. 22, analogous to the burndown charts commonly found in most agile methodologies. The outstanding risk at any point in time is the cumulative risk score (as determined during risk assessment) of all tasks pending completion together with completed tasks resulting in residual risk. Risk is reduced by one or more of the following conditions:

- Completion of risk related tasks. These recognize efforts taken to address the underlying risk (e.g., exploit or reduce).
- Decisions taken to affect a course of action in relation to a risk. These typically involve the reduction of inherent risk to residual risk in acknowledgement of a proposed measure (e.g., transfer or share).
- Decisions taken to perform tasks in a specific manner. This acknowledges the contribution of agile technique to the reduction of risk.
- Cessation of activity that gives rise to a risk. These concede that the risk invoked by the activity is infeasible or uneconomical to bear or share (e.g., avoid).
- Expiry of a risk. Recognition of activity that having been undertaken causes the source of the risk to dissipate.

On account of the fact that the risk effect of agile techniques has been apportioned to individual user stories during risk assessment there is no need to account for it separately again in the risk burndown. The *iteration residual risk* represents that risk which cannot be entirely eliminated by risk management activities and lies somewhere between zero and the sum of all residual risks.[8] Iteration residual risk is a reminder to all project members that risk cannot be entirely eliminated and is a visual depiction of the extent to which unmanaged risk resides in the project. High iteration residual risk is a matter of concern and should be examined in the context of the project and overall enterprise risk appetite. Typically the iteration residual risk consists of:

- Risks associated with accept strategies where it has been decided to rely on a contingency plan in the event that the risk occurs.

---

[8] Neither endpoint seems plausible for a typical project so the actual figure will indeed be somewhere in between.

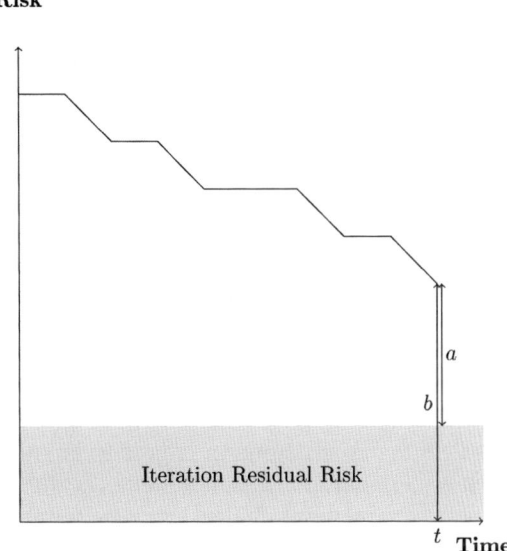

- Residual risk of tasks where completion of a task does not entirely eliminate the risk (e.g., scalability measures to cater for unanticipated demand up a predefined level beyond which the risk of additional demand constitutes residual risk).
- Risk over which there is an element of loss of control arising from the action undertaken that persists beyond completion of the task (e.g., transfer or share strategies where a liability is retained giving rise to residual schedule risk).

As the iteration progresses risk related activity will gradually erode overall risk until it approaches the iteration residual risk (below which no further reduction is possible) or falls short of it (due to the presence of incomplete risk related tasks). No increase in risk will be encountered during this period unless a reassessment of risks forces upward pressure on current risks. If risk assessment is conducted once at the start of the iteration it is probably only safe to conclude that the curve is monotonically decreasing throughout the iteration provided that scope does not change.

A useful risk indicator in this context is the *iteration managed risk ratio* which we define as,

$$\text{Iteration Managed Risk Ratio} = \left(1 - \frac{\text{Iteration Residual Risk}}{\text{Total Iteration Risk (incl. Iteration Residual Risk)}}\right) \times 100,$$

reflecting the proportion (expressed as a percentage) of manageable risk remaining in the iteration. At the start of the iteration this figure will start high (e.g., 85 %) but over time this ought to decrease down as risk is managed away (or systemic risk rises!) until the ratio reaches zero. This, however, does not imply that there is no risk but merely that the only risk remaining is systemic to the iteration and cannot be entirely eliminated. The iteration manaed risk ratio at point, t, in Fig. 22 above, is computed directly by dividing a by b.

A final visit to our website migration example suggests that our DNS configuration task has been created and assigned to an external expert. This decision to deal with the original risk in this manner immediately earns us a deduction of five points from the risk burndown (even prior to completion of the task) since we are now only faced with residual risk (e.g., supplier risk) which itself is eliminated on completion of the task (i.e., a further reduction of one from the risk burndown).

Concluding Remarks

Needless to say the riskier a project is, the more embedded risk management it requires. The question arises, however, at what point do the inherent practices of agility fail to provide sufficient safeguard for projects. Our experience suggests that for small projects tasked with clear and stable targets a light touch approach is adequate (e.g., application of risk management at the incremental level). The mere exercise of risk analysis does appear to inject a sense of realism in a team and discourages a "rose tinted" view of what lies ahead. As project complexity and size grows the need to analyse, track and manage risk rises and there is undoubtedly a case to be made for a better understanding of the project risk environment and its consequences. As we discussed in the Chapter "Project Risk Management" teams that are not co-located (e.g., overseas suppliers) bring new cultural dimensions into play that present acute challenges to agile capabilities and the handling of risk. Research appears to suggest that the integration of risk management is particularly valuable where there is a confluence of several factors (e.g., team size and distribution, project risk exposure and criticality) and that this can be achieved without seriously impeding agile capabilities (Nyfjord 2008).

Applying Agile Risk Management

Abstract Having covered the agile risk management process in the Chapter "Agile Risk Management" we now turn our attention to our three core methodologies (XP, Scrum and DSDM) to illustrate its application. For each methodology we trace its background describing key features in detail, summarize it in terms of the methodological dimensions we introduced in Chapter "Agile Software Development" and interpret its practices using agile charting. We then proceed to describe how agile risk management might be applied to each methodology and refer where appropriate to the Chapter "Agile Risk Management" when making general remarks not specific to any methodology (e.g., the practice of risk scoping).

General Advice

We recommend in general that risk management be addressed as a specific undertaking in the project. We encourage enterprise level identification of risk drivers and that the project be risk scoped within this context during project initiation. We believe that the comparability of risk profiles helps transfer of learning between projects and ensures that project risk is consistent with enterprise risk. We feel that the tools presented in the Chapter "Agile Risk Management" (e.g., risk lists, tagging of risk stories, risk burndown) can be applied to a wide range of methodologies and that risk identification, treatment and monitoring should be integral and conscious parts of project activities.

It is in the bringing together of the main artefacts of the Chapter "Agile Risk Management" that the best expression of transparency and balance can be found. We suggest constructing a risk wall comprising of the following elements (see Fig. 23).

1. *Project Risk Driver Map*. This serves as a reminder of how the project risk fits into the overall organisation and where specific areas of risk can be found.
2. *Project Agile Chart*. This describes the approach to the project but may be adapted as the project progresses (e.g., via feedback from iteration reviews).

Fig. 23 Mock of a risk
wall. Published with kind
permission of © Alan Moran
2013. All Rights Reserved

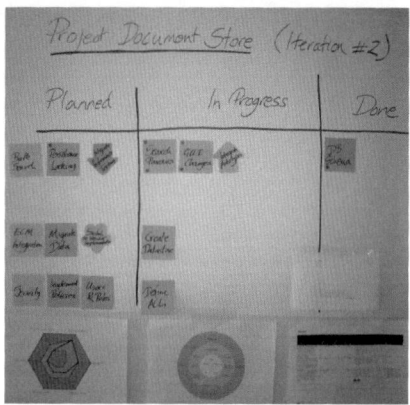

Making it available with the suggestion to annotate opportunities for improvement (e.g., circling elements that need attention) is an excellent way of gathering feedback during an iteration.

3. *Risk List.* New and emerging risks not already captured during the iteration risk workshop can be added by any team for analysis at the next available opportunity lest they be forgotten. This helps promote risk awareness and communication.

4. *Risk modified Kanban board.* This highly visual device ensures that the distribution of risk and its relation to reward (i.e., value of user stories) remains visible to all and helps frame questions about how much risk is being engaged in the pursuit of value. The board should include not only risk tasks and contingency plans but also risk tagging of all tasks.

5. *Risk Burndown.* This communicates how the effectively and efficiently the team is tackling risk as well as how much systemic risk the project is exposed to. This should correlate with the prioritization found on the Kanban board and the parameters of the project risk driver map.

Figure 23 shows how a risk wall might be constructed. The first row of the Kanban board (divided into "Planning", "In Progress" and "Done") illustrates a user story with several tasks amongst which risk reduction (downward arrow) and a risk exploitation (upward arrow) tasks can be found. Several tasks are also tagged with coloured circles which are identified in the risk list legend (see bottom right). Of the remaining user stories only the second has a risk task which in this case in a contingency (the cloud sticker) relating to an accepted risk on the risk list. Lining the bottom of the risk modified Kanban board are (from left to right) the project risk driver map, the project agile chart and the risk list above which the risk burndown chart can be found. At this point were someone to walk into the room, (s)he could instantly understand which stories bear the most risk, how these risks are being tackled and evaluate accurately the overall riskiness of the project.

Fig. 24 XP methodological dimensions. Published with kind permission of © Alan Moran 2013. All Rights Reserved

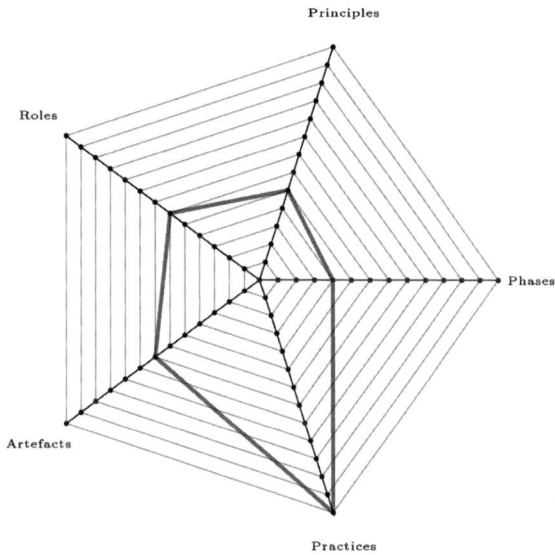

eXtreme Programming

In light of research specifically relating to eXtreme Programming (XP) practices our critical assessment of its attitude toward risk concludes that whilst the engineering focus of XP might permit it to exploit risks in a flexible manner, there may be structural limitations to the extent to which it is capable of deploying risk response strategies. We nonetheless identify some specific practices from which XP might benefit and highlight the issues with XP that could warrant management attention.

Methodology Overview

Extreme programming was founded by Kent Beck, Ward Cunningham and Ron Jeffries and grew out of the efforts, whilst working at Chrysler in the 1990s, to develop a methodology based on existing best practices taken to the extreme. The result was a rich interwoven set of techniques that reinforced each other and today define the essence of an agile software development infrastructure. XP is a high discipline methodology that requires adherence to standards and a strong commitment to unit testing, refactoring and integration. The lack of focus on document output that characterises XP teams is an aggressive expression of the "working software over comprehensive documentation" statement in the agile manifesto. Indeed XP eschews most forms of administrative overhead (including reporting) and focuses entirely on engineering practices. In general XP advocates generalists who can contribute to many facets of a project, work closely together and share their knowledge.

Figure 24 depicts an overview of the XP methodology in terms of the basic dimensional parameters introduced in the Chapter "Agile Software Development" which

Table 20 XP methodological dimensions

Dimension	Description
Principles[a]	Simplicity, Communication, Feedback, Respect, Courage
Roles	Customer, Analyst, Programmer, Tester, Coach, Manager
Artefacts	User Stories, Release Plan, Iteration Plan, Acceptance Tests, Documented Coding Standards, Vision (derived from System Metaphor), (Architectural) Spikes
Practices	Whole Team (incl. on site presence of the customer), Planning Game, Small Releases (i.e., incremental delivery), Customer Tests (pref. automated acceptance testing), Simple Design, Pair Programming, Test Driven Development, Design Improvement (i.e., refactoring), Continuous Integration, Collective Code Ownership, Coding Standards, System Metaphor (i.e., Vision), Sustainable Pace
Phase	Release Planning, Iteration, Acceptance Testing, Release

[a]XP refers to these as values

we have reinterpreted in the interests of comparability (see Table 20). This confirms the technical focus of XP and whilst opinions concerning the precise nature of the other dimensions vary according to source (Jeffries 2013; Wells 2009a; Beck and Andres 2004), there is a broad consensus concerning what constitutes XP and what membership of an XP team might entail. In particular XP is less model theoretic when compared with Scrum or DSDM and thus our phases are inferred and constructed from descriptions of how practices are employed (Wells 2000).

The attitude of XP towards roles[1] is slightly ambiguous. On the one hand it is clear that a skilled and disciplined practitioner is necessary, whilst on the other hand the notion of a "generalizing specialist", coined in (Ambler 2003), is expected by virtue of the practices of pair programming and moving people around in teams. This is a balancing act that requires skill though it reflects a wider sentiment in the community (Boehm and Turner 2009). Roles are, however from a risk perspective important, since they concern people risk (discussed later) and the ability to manage it.

Agile Charting

Let us now consider how XP practices can be interpreted in terms of an agile chart as introduced in the "Agile Software Development" chapter. Owing to the fluid nature of XP, our interpretation, depicted in Fig. 25, should be understood as indicative rather than normative.

At first glance the distinction between increments, referred to as releases, and iterations (occasionally referred to as "sprints") is somewhat optional in XP as evidenced by the remark

> Some Agile projects don't even have iteration ends. They remain Agile by balancing the need for stable requirements with the need to change requirements while launching working software on a regular and frequent basis. (Wells 2009b)

adding that

---

[1] Our particular list of roles was sourced from (Jeffries 2001).

Fig. 25 XP agile chart. Published with kind permission of © Alan Moran 2013. All Rights Reserved

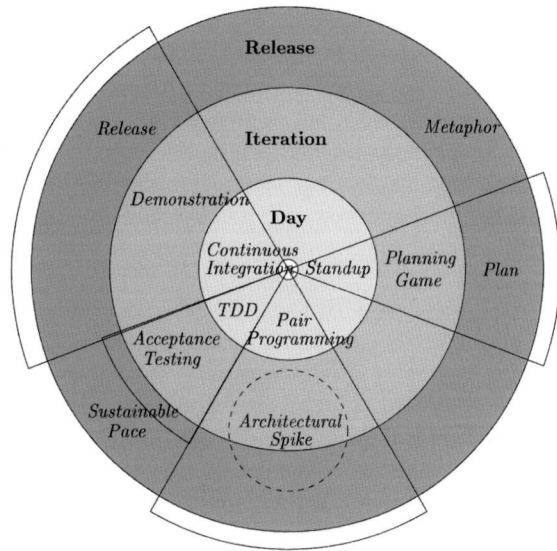

iterations are just an easy way to demarcate when changes are accepted, when a new plan is created, and when working software is released to customers (Wells 2009b)

On account, however, of the separation of release and iteration planning (Wells 1999b) we choose to retain the distinction in our charting of XP. Releases are thus characterized as being where the vision (i.e., system metaphor) and team composition are determined (if these are not already agreed in the pre-project phase) and the point when high level bundling of features together with the planning of their release (and hence the definition of sustainable pace) occurs.

This leaves the iteration cycle concerned with the details of user stories together with their prioritization and estimation all of which takes place during the planning game. It becomes necessary to augment this with modelling and prototyping which XP refers to as spikes (though at a higher level some consensus surrounding architectural aspects might have been decided at the release level). Acceptance testing that occurs at this level involves the customer directly and supplements the unit test testing that occurs on a daily basis. This cycle concludes with a demonstration of working software at the end of the iteration.

Daily activities revolve around the stand-ups at the start of each day (themselves a form of planning exercise) and the concrete practice of code ownership (e.g., pair programming) and validation of production (e.g., test driven development and continuous integration) which ensure that feedback is engrained in the process at a very deep level. The concept of daily build and deployments is seldom explicitly cited in XP though it could be argued that it is present to a certain extent in continuous integration.

We have not cited some elements of the XP framework in this description. For example it is possible that coding standards are best elaborated at the release level, in the pre-project phase or at the corporate level where the concept of collective ownership could also be supported. Moreover whilst we feel retrospectives are an appropriate exercise at the end of each iteration, XP is not prescriptive about when these should occur. Finally simple design represents a guiding principle that ought to be elucidated at the iteration level and practiced on the daily level. All of these are, however, merely comments and it is perhaps best that the team decides for itself when and how often such practices should be engaged in.

XP exhibits natural slicing in respect of planning, review, testing and deployment (each illustrated separately in Fig. 25). For example when deciding in a specific project context at what level certain types of review exercises should occur (e.g., an innovative design idea) it is worth focusing on pair programming and (architectural) spiking to consider possible trade-offs. Equally the deployment process is well supported at all levels all of which clearly demonstrate working software leaving flexible the option of where to embed specific deployment related activities (e.g., deployment of built artefacts into central corporate environments or practice of continuous delivery). Similar remarks apply to the planning and testing slices.

Risk Scoping

XP addresses risk by accepting the underlying uncertainty of customer requirements and ensuring that the development process anticipates and is capable of coping with change. Indeed XP was specifically conceived to "address the problems of project risk" with practices that "are set up to mitigate the risk and increase the likelihood of success" (Wells 1999c). This notwithstanding, there are no practices specifically focused on the identification, analysis and management of risk nor can the methodology be said to be risk-driven in the sense of (Garland and Fairbanks 2010). Indeed the extreme emphasis on low ceremony and the disinclination towards heavy documentation might even inhibit the adoption of the more detailed analyses that would otherwise be recommended by risk-driven approaches. Our analysis of the archetypal risk drivers identified in this Chapter and how XP tackles them is presented in Table 21 and assumes that the practices of XP occur within some form of project context (which of course is not always the case).

Risk Tailoring

We suggest that risk assessment take place as part of the iteration planning at the start of each iteration. According to XP this is the point at which sufficient information is available to make detailed estimates and ought therefore to be risk assessable. We feel that risk management at release level is likely to be less effective for this reason and thus perceive practical issues with the approach advocated in (Li et al. 2006) as described in the Chapter "Agile Software Development" (Fig. 26).

Table 21 Assessment of archetypal risk drivers for XP projects

Risk Driver	Description
Technical risk	There is a pervading perception of negative risk that is reduced through the use of (architectural) spikes and sharing of knowledge through collective ownership practices such as pair programming. Clearly simple design and refactoring also play a role here in the reduction of risks
Requirements risk	Similar focus on negative risks found in technical risk with technocratic countermeasures in the form of user stories augmented by test driven development and continuous integration which validate and detect regressional errors and unwanted side effects
Schedule risk	Treatment appears to be influenced by lean operations management where the notion of small batch sizes i.e., frequent small releases, takes centre stage. Sustainable pace is an admirable trait that could contribute to risk containment if properly managed
Project risk	The endorsement of retrospectives ("Fix XP when it breaks" (Wells 1999a)) is an expression of the handling of project risk that is commonly found in agile methodologies. Interestingly it is seldom formulated in positive terms (e.g., "extend XP to make it better") though this may be a defence of methodological integrity
Supplier risk	XP possesses few coping mechanisms for tackling supplier risk. Indeed its requirements that the team be co-located and that it have immediate and continual access to the customer representative (themselves risk mitigation measures) may even constitute a limitation in this respect if the supplier cannot comply
People risk	The assumption of skilled and disciplined team members translates into a requirement that exposes a risk if not satisfied. Notwithstanding practices such as pair programming and moving members around (part of the code ownership portfolio of measures) can reduce risk by building knowledge and skills. Sustainable pace also ensures that this risk is not exaserbated by project risks

Risk tailoring in the context of XP advocates the design explorative practice (e.g., prototyping or architectural spikes) early in the iteration reinforced by refactoring and involvement of customer representatives in order to manage requirements and technical risk. Moreover the agility of XP provides ample room for exploitation of upside risk were awareness thereof promoted. Iteration planning ought to include a risk identification and analysis workshop to determine appropriate risk tasks for the iteration. Supporting this in the context of high risk projects is the XP agile infrastructure (i.e., continuous integration, dedicated build server, automated testing) that is likely to prove significant support for managing change and uncertainty in the code.

Noting the observations already discussed in (Li et al. 2006), we suggest that a risk evaluation at the release cycle level might be appropriate but only in so far as this makes clear to the customer what the risks of a release entail. This is akin to aggregation of risks as described in (ISACA 2009b) and should serve to comment on the overall level of risk rather than be a basis for a detailed risk analysis and management which we reiterate should occur at the iteration level. The fluidity of XP makes it an excellent methodology to practice escape velocities (see the Chapter "Agile Software Development") and it is here that risk related conditions could be set (e.g., escape criteria based on the iteration residual risk ratio).

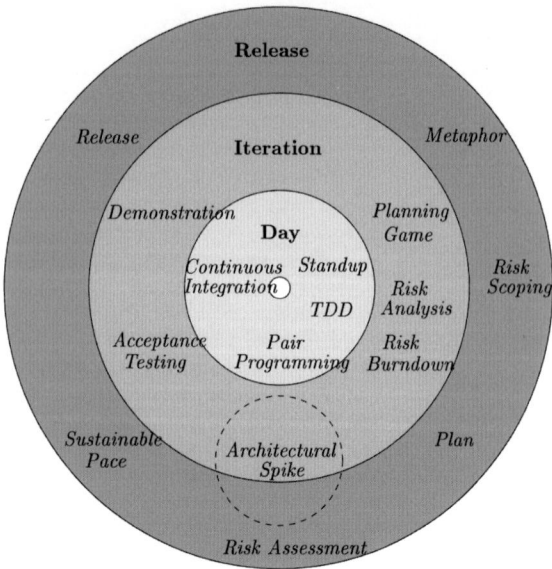

Risk Management

The natural point at which risks can be identified is during the planning game when
user stories are being formulated. Indeed task risks can be linked to user stories and
thus constitute user stories in their own right in which either a specific risk is analysed
or managed. One of the rare explicit pieces of advice concerning risk proffered by
XP takes up this point when it reduces a technical risk to a question of estimation
and cites the following technical uncertainty.

> Well, if C# has the same kind of morphilation facilities as Java, then this will be a two. But
> we don't know if it does or not. If we have to write a morphilation library, this card could
> be a thirty or more. (Jeffries 2006)

It correctly argues that there is little value in estimating such a task but rather that
there is more value to the customer in creating a new task that settles the matter.
In effect what is being advocated here is a risk analysis activity. A risk manager
might have tackled this matter slightly differently beginning with an understanding
of the risk environment and appetite of the project. This might, for example, raise
the question as to whether the use of "morphilation facilities" constitutes a highly
innovative step in an otherwise risk averse project. Such matters cannot simply be
swept aside since the project sponsor has an obligation to manage risk that falls
under the remit of project governance. The first activity would be to identify the risk
perhaps using a risk statement of the form

> As a result of the need for Java-like morphilation facilities, there is uncertainty as to whether
> or not our chosen programming language, C#, can provide these which could force us to
> arrive at an alternative thus adversely impacting our timeline.

We choose to focus here on the schedule risk (though the framing of the original statement also highlights an implicit technical risk). Based on our assessment of likelihood and impact we might be presented with risk responses of avoid (try not to rely on morphilation), reduce (develop own morphilation library) or transfer (purchase an existing morphilation library). If we transfer then we might simply be exchanging technical risk for (a reduced) supplier risk. Either way our response requires a residual risk assessment to decide if we are on the right path. Risk is therefore a management activity (albeit performed by anyone in the team) that is distinct from estimation and other activities.[2]

Risk estimates arising from analysis should be a part of the project velocity (a measure of how much work is being done in an iteration derived from the sum of user story estimates) and must therefore be taken into consideration when setting the sustainable pace of the project. Risk prioritization must reflect the risk exposure and priorities of the team. For example risk might be prioritized according to their relationship to user story priorities (i.e., high priority stories have their related risks accordingly prioritized) or may be based on the maximisation of risk reduction (i.e., risk stories prioritized in descending order of risk estimate) though this may be at the cost of maximization of customer value. The balancing act lies in finding a common prioritization between user and risk stories which in turn relies on the judgement of the entire team.

Table 22 describes how the traditional response strategies relate to XP. The risk pyramid (see the Chapter "Agile Risk Management") appears to suggest that XP might have difficulty coping with low likelihood extreme threats and opportunities since these are likely to be linked to sharing and transfer response strategies which may inhibit specific core practices (e.g., pair programming) and assumptions (e.g., co-location of the team). Substitute responses (e.g., avoidance) might therefore have to be resorted to.

On a positive note there is considerable scope for exploitation of opportunities in XP and the flexibility of the methodology welcomes their discussion and incorporation. This inevitably impacts on planning and here caution is called for that requirements and technical risk management is not giving birth to schedule and project risk.

Conclusions

Of the three methodologies under consideration in this book, XP offers the least scope for tackling risks or embedding risk management practices. Notwithstanding, hope lies in the capacity of XP to pursue opportunities as positive risks where its flexibility offers it considerable advantages over other methodologies. That said, we feel that it is nonetheless structurally constrained in respect of handling some risk strategies (e.g., sharing and transferring) and that its culture encourages it to avoid structuring

---

[2] This should in no way be understood to detract from the original point being made with this example that an analysis to uncover the source of risk is in itself a task of value to the customer!

Table 22 XP interpretation of the classical risk strategies

Risk strategy	Application
Accept	This appears to be the default strategy for schedule risk as evidenced by the quip that "when you slip your schedule you acknowledge how far you have come and make a new plan based on the new information" (Wells 2009b). Otherwise the strategy is seldom mentioned
Exploit	This response is rarely cited so there exists the opportunity to raise the awareness of this strategy and to promote its use
Share	This strategy might be inhibited by the nature of XP and may increase other risks. For example, if sharing means that pair programming becomes more challenging then this may impede the handling of technical risk
Transfer	This can be a challenging strategy to implement depending on the extent to which there is integration of the risk bearers within the project team. Few practices in XP lend themselves to collaboration outside of the team and many place constraints on how externals may participate. It is likely to be fair to conclude that XP has not adequately anticipated this risk strategy
Reduce	This is the default response for most of the risks cited by XP and especially those in relation to technical and requirements risk
Avoid	Though seldom explicitly discussed in the XP literature there is evidence of engagement with the assessment of innovative solutions (e.g., spikes, modelling, communication) which one must conclude does not preclude avoidance

and scoring of tasks in such a manner as to enable more effective risk management and reporting. This is not to say that XP does not acknowledge or tackle risks in its own way, merely that the issues are narrowly framed. Inevitably any undertakings to risk optimize XP may come at the cost of the methodological integrity which might be unacceptable to purists.

Scrum

We survey Scrum as a framework rather than a process and acknowledge the limits of our analysis when conducted outside the circumstances of a specific project environment. This notwithstanding we recognise the same implicit and passive understanding of risk management found in many other agile methodologies but acknowledge the scope for and promise of embedding risk management activities many of which fit well into the framework.

Methodology Overview

Scrum owes it origins to (Takeuchi and Nonaka 2013) where the term was first defined though it fell to Jeff Sutherland and Ken Schwaber to refine and popularize it. Based on empirical process control its practices are founded upon notions of inspection and adaption reminiscent of lean thinking. Scrum can best be described as a product development methodology with project management aspirations as its focus lies in

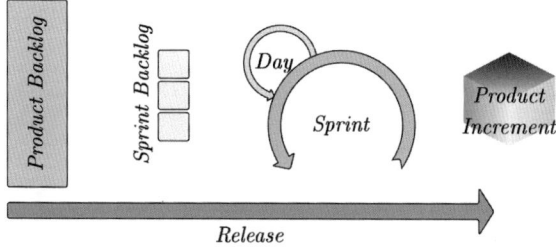

Fig. 27 The scrum process model. Published with kind permission of © Alan Moran 2013. All Rights Reserved

Table 23 Scrum methodological dimensions

Dimension	Description
Principles	Focus, Courage, Openness, Commitment, Respect
Roles	Product Owner, Scrum Master, Development Team
Artefacts	Product Backlog, Sprint Backlog, Iteration
Practices	Sprint, Sprint Planning, Daily Scrum, Sprint Review, Sprint Retrospective
Phases	Sprint, Release

the management and reporting of software requirements and development. The scope of Scrum does not reach to other IT activities such as systems development or migration and it defers to existing practices within an organisation to cover project initiation, risk management, release and deployment and change management processes. Indeed agile methodologies that have a wider scope have argued that Scrum can be successfully embedded within their frameworks (DSDM Consortium 2012b; Craddock 2013; IBM 2011). Scrum shares a common heritage with XP and both employ similar terminology and practices. Differences are, however, apparent in their structure and philosophy. For example, whilst XP focuses on engineering practices, Scrum structures the development process (see Fig. 27) and is less prescriptive on matters of technique (though it does impose a variety of process constraints concerning team size, iteration duration etc.).

Scrum is considerably understated in terms of methodological dimensions and though we remain faithful to the mainstream literature (see Table 23), we note that additional elements are often necessary. For example, Scrum designates as a developer[3] anyone who is neither a Product Owner nor a Scrum Master "regardless of the work being performed" (Sutherland and Schwaber 2013). Other methodologies differentiate roles more finely and it is to be assumed that most Scrum teams do so in practice. Figure 28 illustrates the apparent lightness of Scrum based on the summative information found in Table 23.

As will become evident, Scrum focuses very much on providing a framework for development practices along with a means for reporting and monitoring progress.

---

[3] Sometimes the function designation "The Team" is used.

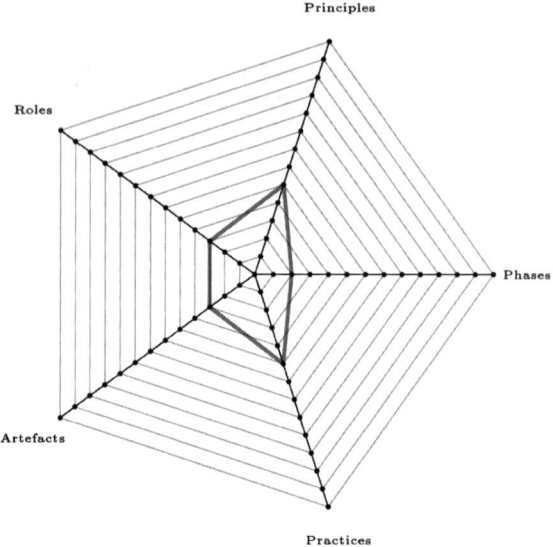

This added structure comes at the cost of the flexibility afforded by XP but replaces it with the freedom to adopt whatever techniques the team deems appropriate. Note that in Table 23 Scrum refers to values rather than principles and events rather than practices.

Agile Charting

Figure 29 depicts our interpretation of a Scrum agile chart based on multiple authoritative sources (Sutherland and Schwaber 2013; Schwaber 2004; Pichler 2010). The distinction between increment (referred to as a release) and iteration (referred to as a sprint) is not very clear in Scrum though like XP there is a planning distinction between product and sprint backlogs which can be used to define the boundaries between the two. Thus it is at the release level (if not at the pre-project level) that the product vision is established and the coarse grained planning, in the form of epics recorded in the product backlog, occurs.

Sprints begin with a planning session which may employ planning poker, a simplified form of the Wideband-Delphi technique, and result in the creation of estimated and prioritized user stories grouped into epic categories that were created at the release planning level. User stories are assigned story points as described in (Cohn 2010) and the project velocity expressed in terms of points per Sprint, which given the fixed timebox nature of an iteration indirectly translates into a point-to-day exchange rate determined by the slope of the burndown chart (which graphically depicts the remaining points as a function of time). Completion criteria are expressed as definitions of done that can be applied to the story and sprint levels. Sprint productive work concludes with a review to "foster collaboration and elicit feedback"

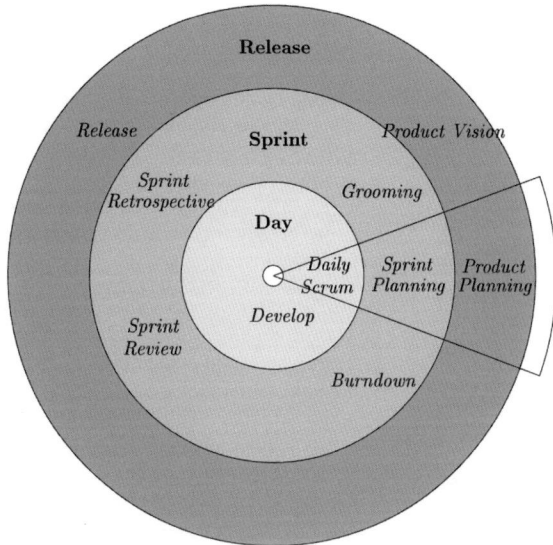

Fig. 29 Scrum agile chart.
Published with kind permission of © Alan Moran 2013.
All Rights Reserved

(Sutherland and Schwaber 2013) which, despite claims to the contrary, has the character of a status gathering meeting. It is during the review that the work completed in the Sprint is demonstrated. Note that it is permitted to challenge the value proposition of the project as well as the contents of the product backlog in this meeting. The review is followed by a retrospective which endeavours to identify means to improve the process, a practice which together with Scrums and Scrum-bans[4] illustrates the transparency that Scrum promotes. Feedback should be planned and integrated in time for the next Sprint.

Each day starts with a structured stand-up meeting, referred to as a Scrum, wherein progress since the last meeting, plans for the day and impediments are discussed. Team members are understood as the primary active participants though other passive participants may also be permitted to attend.

Scrum has only one natural slice in the form of planning though others are likely to become apparent within specific projects once their practices are decided upon. The lack of slices at this level simply reflects the framework character of Scrum.

Risk Scoping

To a certain extent Scrum considers not only risks that arise in projects but also process risks noting the ability of the process itself to compensate for undesirable variances. Although Scrum explicitly cites the need to be aware of process deviation "outside acceptable limits" it fails to define where such limits are set nor does it offer tracking tools to monitor them. This, of course, is the function of risk scoping wherein

---

[4] Scrum-ban is an adaptation of Scrum which includes a tracking practice borrowed from Kanban.

Table 24 Assessment of archetypal risk drivers for Scrum projects

Risk driver	Description
Technical risk	Whilst not specifying precisely which tools to employ the principles of transparency, learning and feedback applied to technical matters help reduce risks
Requirements risk	The use of user stories with definitions of done achieve a high degree of clarity in a language that the customer understands. Feedback loops and other adaptive practices (e.g., ability to clarify and renegotiate user stories) together with the inclusion of the customer in planning and review meetings also helps keep these risks in check
Schedule risk	It is argued that development horizons are correlated to complexity which in turn implies risk (Sutherland and Schwaber 2013). Thus the short iterations (not exceeding one month) are claimed to reduce risk and cap financial risk both of which ought to reduce schedule risk
Project risk	The empirical process control principles atop of which Scrum is built are intended to detect unwanted variances and to adapt plans accordingly. This reflects elements of corrective action found in controls based risk management
Supplier risk	Relatively little attention is given to supplier risk. The belief in scalability of Scrum expressed through the concept of Scrum of Scrums would be a situation where intra-Scrum supplier risk might arise but here too there is little to be found here in the literature. Accordingly we judge this to be a largely neglected aspect of risk management
People risk	The Scrum Master is tasked with multiple supportive roles (e.g., coach, facilitator) which aid in the handling of risks. Otherwise its process structure might be said to make it easier to identify people risks (since the expectations placed on them become clearer) even if Scrum says little about how they should be resolved

appetite and tolerances are discussed within the team and it seems conceivable that these could be embedded into Scrum without any adverse impact on agile capability. Moreover there is no need to handle project and process risks differently (e.g., in terms of using risk registers and pyramids) and thus such risks could continue to be treated as one might expect. Table 24 presents the attitude of Scrum towards the generic IT risk drivers already identified in the Chapter "Agile Risk Management".

No explicit role assignment as risk manager is mentioned in Scrum though the related notion of "responsibility for maximizing the value of the product" (Sutherland and Schwaber 2013, p. 5) is assigned to the Product Owner. This notwithstanding it may well fall to the Scrum Master to ensure that operational aspects of risk are pursued and that the principles of risk management are adhered to. Feedback based on inspection and adaptation pervades Scrum at all levels and offers a means of reducing risk when understood as cognitive uncertainty.

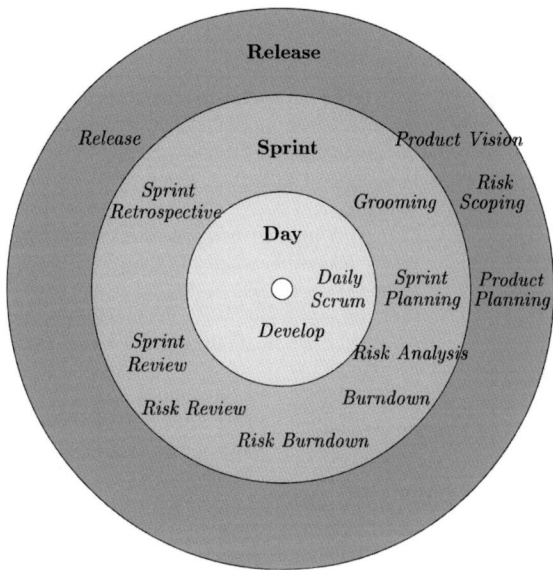

Fig. 30 Risk adjusted Scrum agile chart (slicing omitted). Published with kind permission of © Alan Moran 2013. All Rights Reserved

Risk Tailoring

Figure 30 depicts the risk tailoring of Scrum which serves as a basis for more project specific tailoring.

Risk scoping should be assessed at the start of every release cycle to revalidate the premise on which the project is based. However, it is at the Sprint level that risk analysis (including identification and prioritization) should occur alongside risk reporting that builds upon existing burndown practices as described in the Chapter "Agile Risk Management". The retrospective provides a forum to determine if risk measures have been implemented and were effective. Given the empirical process focus of Scrum the retrospective (and review) also offer the opportunity to discuss process risk and how the next Sprint needs to be amended.

Finally since it is not unlikely that XP practices (e.g., continuous integration, refactoring, pair programming) are to be found within the Sprint or daily cycles the remarks made in relation to XP also apply here with the proviso that the structural constraints of Scrum (e.g., not altering the backlog scope or quality specifications) are not violated.

Risk Management

As with other methodologies the natural point of risk identification, analysis and prioritization is at the Sprint level and similar remarks apply here as already cited

Table 25 Scrum interpretation of the classical risk strategies

Risk strategy	Application
Accept	It might be assumed that this would be inherent in Scrum thinking though it is infrequently cited and the link between choice of this strategy and risk exposure is essentially non-existent suggesting that it is at best an implicit recognition
Exploit	In keeping with the consensus of the community, opportunities in projects are seldom cast as positive risk to be exploited
Share	Compared with XP there is a stronger belief in the ability to share risks and though this is seldom mentioned in the basic form of Scrum it could be understood to feature more strongly in enterprise notions of Scrum
Transfer	Few explicit references are made to risk transfer suggesting that it is not consciously considered in Scrum or that awareness thereof is not widely present
Reduce	Principles of transparency inspection and adaption (Sutherland and Schwaber 2013) underpin efforts to become aware of and tackle risks and reduced unwanted process variations
Avoid	The reluctance to introduce changes that endanger Spring goals could obliquely be understood as a tendency towards avoidance of risk generating activities. Otherwise this strategy is seldom mentioned

for XP and described in this Chapter. In particular the tension between risk minimization and value maximization arises here too and it should be borne in mind that both do not necessarily pull in the same direction. The structured user story basis of Scrum product planning together with the framework approach that admits different techniques to be employed both make possible the integration of agile risk management as conceived in this book. Rather like XP, the integration of risk stories, scoring and tagging and burndown charting ought not pose a challenge for Scrum. Table 25 describes how the traditional risk response strategies might be interpreted in a Scrum context.

Conclusions

Scrum's framework character leaves open a more detailed analysis of its risk management capacity. What can be said, however, is that its awareness is implicit and therefore its capability meagre suggesting that augmenting it with explicit risk management practices is to be recommended. The scope for such activity is both broad and potentially fruitful giving hope to managers and practitioners alike that risk optimisation is possible. The jury is still out on the linkage of enterprise Scrum (e.g., Scrum of Scrums) to the wider context of enterprise risk management though we return to this topic in the Chapter "Enterprise Agility".

Fig. 31 The DSDM process model. © DSDM Consortium 2013. Reprinted with Permission

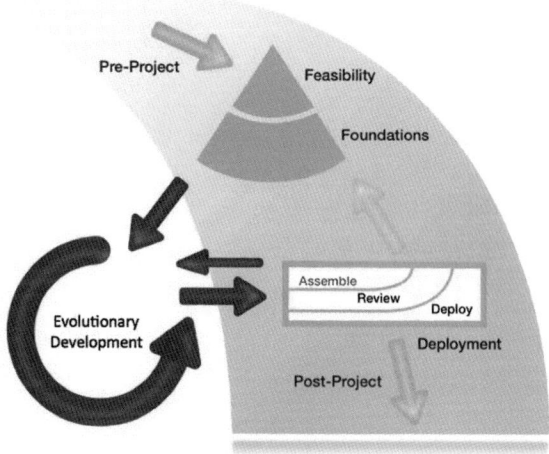

Dynamic Systems Development Method

Methodology Overview

DSDM as understood throughout this book has often included the agile project framework, AgilePM®. In this section, however, we place more emphasis on the underlying model of DSDM and note that there are some methodological differences between this and the current version of AgilePM®. The six phase model that we describe is endowed with a rich set of roles and artefacts making it suitable for corporate project environments with established portfolio management and governance practices in place that may be willing to embrace agility but who seek reassurance that control can still be retained. Indeed the manner in which projects are structured ought to be reasonably familiar to those working with traditional project management methodologies (e.g., PMI, PRINCE2®). The DSDM model, depicted in Fig. 31, stipulates a feasibility and foundations phase followed by an evolutionary development[5] activities that culminate in one ore more deployments.

DSDM grew out of efforts to make more agile Rapid Application Development (RAD) which rose to prominence in the 1990s. It has positioned itself as a project focused framework on occasion releasing advice on how to embed other methodologies (principally XP and Scrum). Rather unusually for an agile methodology, it embraces a very wide spectrum of activities throughout the project lifecycle explicitly citing topics such as quality, risk and configuration management. Accordingly it is rather more likely to appeal to those in a mature corporate environment who are

---

[5] Earlier versions decomposed this phase into Exploration and Engineering phases.

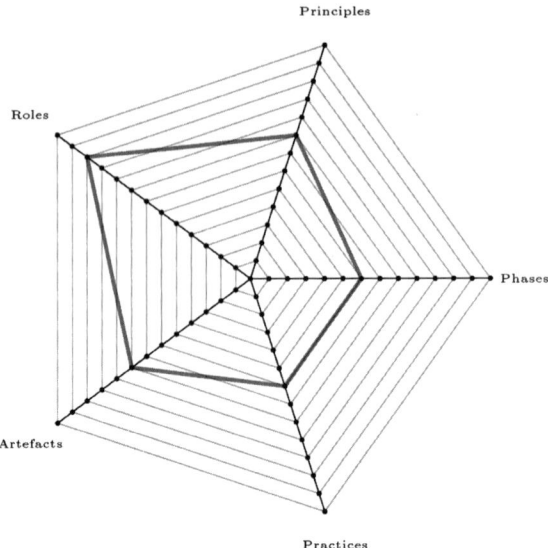

Fig. 32 DSDM methodological dimensions. Published with kind permission of © Alan Moran 2013. All Rights Reserved

already accustomed to the weight of ceremony[6] implied by the methodology as well as those seeking to link agility to governance and agile portfolio management (DSDM Consortium 2012a). As we alluded to in the Chapter "Agile Software Development", this weight of DSDM is due in part to the requirement that it expresses compatibility with lesser agile industry frameworks, a decision that was consciously taken at its inception in 1994. Whilst the number of artefacts may appear overwhelming to most agilists, DSDM is at pains to stress that these do not constitute full documents in the traditional sense stating that

> Substituting traditional 'big design up front' with DSDM's 'enough design up front' promotes Agility in developing the required solution whilst avoiding the risk 'no design up front' that makes many larger and more strongly governed organisations so nervous. (DSDM 2012)

At a glance it is clear from Fig. 32, which is based on Table 26, that DSDM is very detailed in the guidance it offers. Note that the additional roles of Workshop Facilitator, Coach and other specialists such as Operations Co-ordinator are considered auxiliary roles in DSDM and are therefore not included in this list. Moreover for reasons of comparability with other methodologies some artefacts are not cited in this list (e.g., Evolving Solution, Deployed Solution) consistent with our other classifications.

---

[6] Though DSDM is rich in advice and artefacts it is important to note that not everything is considered mandatory.

Table 26 DSDM methodological dimensions

Dimension	Description
Principles	Focus on the Business Need, Deliver on Time, Collaborate, Never Compromise Quality, Build Incrementally from Firm Foundations, Develop Iteratively, Communicate Continuous and Clearly, Demonstrate Control
Roles	Business Sponsor, Business Visionary, Project Manager, Technical Coordinator, Team Lead, Business Ambassador, Business Analyst, Business Advisor, Technical Advisor, Solution Developer, Solution Tester
Artefacts	Terms of Reference, Business Foundations (incl. Business Case and Prioritized Requirements List), Solution Foundations (incl. Solution Architecture Definition and Development Approach Definition), Management Foundations (incl. Delivery Plan, Project Approach Questionnaire), Delivery Control (incl. Timebox Plan, Timebox Review Record), Quality Assurance (incl. Technical Quality Review Records, Business Quality Review Records), Project Review and Benefits Assessment
Practices	Facilitated Workshops, Incremental Delivery, Timeboxing, Iterative Development, MoSCoW Prioritization, Modelling and Prototyping
Phases	Pre-Project, Feasibility, Foundations, Evolutionary Development, Deployment, Post-Project

Agile Charting

DSDM describes the development process in terms of cyclical execution of activities that lead iteratively to the final solution and structures the iteration cycle, which it refers to as a timebox, to act as a control mechanism. Thus much of the emphasis of the model is on demonstration of control and providing a comprehensive structure around project activities as can be seen from Fig. 33.

An increment is characterized by the delivery of a deployment unit[7] and involves planning activities leading to the deployment phase comprising of assembly, review and the actual deployment itself. Increments are created by one or more timeboxes which comprise of investigative, refining and consolidatory stages, which DSDM refers to as iterations, sandwiched between short kick-off and close-out sessions (see Fig. 34). Note the use of the term iteration in this context which differs from our use. DSDM uses the term iteration to refer to a structured phase within a timebox, divided into identification, planning, evolution and review stages during which the actual work is done and the outcome is used to decide if another iteration is required. This advice found in (DSDM 2010) is absent from other shorter publications (DSDM 2012) hinting that it is not entirely mandatory. For all intents and purposes it is a timebox that should be equated to an iteration as used in this work. For example, a timebox is comparable with a Sprint in Scrum as indicated by (Craddock 2013) and is generally expected to be between two and four weeks with durations of more than six week discouraged (see Fig. 34). A significant element of timeboxing is the exercise of control and may involve additional practices such as auditing activity in

---

[7] Though whether or not this constitutes a release is left open.

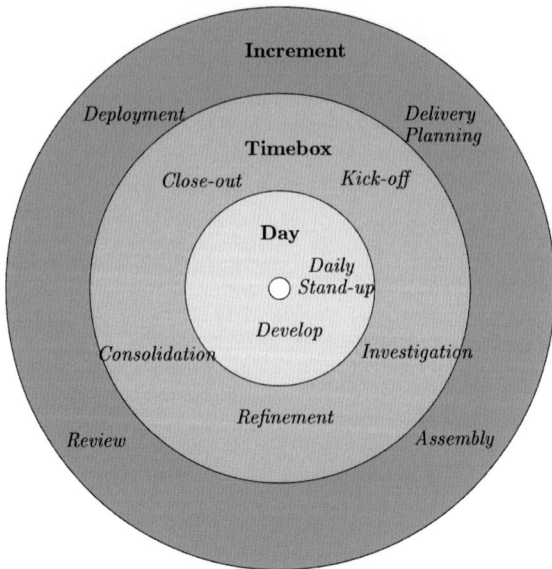

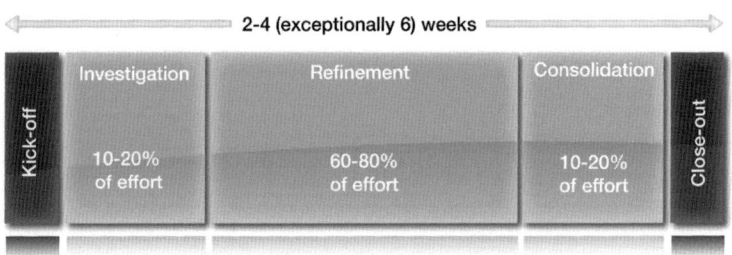

solution review records. Whilst this does appear to be rather heavy practice for an agile methodology it may be justified in some contexts (e.g., compliance within a regulatory environment) and thus this can be said to be an extension of agility to domains more commonly associated with traditional project management methodologies. Beyond occasionally mentioning a stand-up meeting (DSDM 2010) and development efforts, DSDM is rather silent about structure at the daily cycle. This might be attributable to the fact that DSDM sees itself as providing the necessary framework for oversight and project management without stipulating the precise nature of production activities. It is in this respect that we see one of the sharpest contrasts between XP and DSDM both of which we argue occupy different branches of the product-project dichotomy.

Owing to the high level management focus of the DSDM framework we are somewhat reluctant to slice Fig. 33 since this might superimpose on a generic model our presumptions concerning the recursive character of productive activities. This does not preclude slicing in a specific project context where additional tasks would further clarify how such concerns (e.g., quality, testing and deployment) are being concretely addressed.

Risk Scoping

Although we noted already that DSDM frames risk in a purely negative fashion, it does appear to be one of the more risk aware methodologies. Indeed its own materials provide templates based on traditional risk management albeit without any guidance on how to manage risks in an agile manner. Moreover there does appear to be some inclination towards risk scoping through the use of a project approach questionnaire "to identify risks to the agile process" (DSDM 2010) though the structure of this document which is included as an appendix in (DSDM 2010) is somewhat broader than this statement would suggest and endeavours to cover the needs of negotiation with stakeholders, risk managers and auditors. Table 27 indicates how DSDM tackles the main risk drivers we identified in the Chapter "Agile Risk Management" and is based largely on (DSDM 2010) and derivative works.

DSDM has issued a whitepaper on risk management in which a comprehensive range of risks are cited that encompass the above categories and augments this with specific guidance on how to tackle them (DSDM 2003c). This reflects in large part the controlled environment in which DSDM tends to find favour but is nonetheless unusual amongst agile methodologies in terms of its level of detail.

Risk Tailoring

The adaptation of DSDM along the lines described in the Chapter "Agile Risk Management" is unlikely to impose much on DSDM. For example, the inclusion of assessment of risk appetite and determination of tolerances is likely to appeal to the target audience of the methodology though encouraging risk managers to make agile their existing practices is probably more a cultural than a technical challenge. Matters of risk scoping need to be tackled in the Feasibility or Foundations phases where at the very least appetite and risk drivers need to be identified. This notwithstanding any increment must be at least cognizant of the key risks it faces and thus a risk scoping activity at that level would be advisable. As with other methodologies the most effective point in time to commence with operational risk management is within the timebox cycle. Indeed this is precisely what DSDM recommends when it suggests that the solution team review risks at the end of the investigation phase and correlate these against the delivery plan and risk log (DSDM 2013b). Figure 35 suggests an appropriate placement of these tasks on the DSDM agile chart.

Table 27 Assessment of archetypal risk drivers for DSDM projects

Risk Driver	Description
Technical risk	The primary instrument for tackling this form of risk is of course prototyping though communication and the other measures cited for management of requirements risk are also relevant in this context. Note too that the system architecture definition artefact (of the solutions foundation) and the Technical Co-ordinator role also contribute to risk management in this respect
Requirements risk	There is an acceptance of the inevitability of uncertainty in relation to requirements or instability in respect of their specifications. DSDM proposes several mechanisms for tackling such matters (e.g., facilitated workshops, modelling, prototyping, timeboxing) which suggests that such risks register clearly on the radar. Moreover there are several roles (i.e., Business Ambassador, Business Advisor and Business Analyst) involved in clarification and validation of requirements
Schedule risk	The use of contingency in planning (i.e., the distribution of tasks of varying priorities) and timeboxing is argued to be effective means of tackling late delivery (DSDM 2010). There are echoes of schedule risk management in the principle that promises on-time delivery for reasons relating to return on investment and risk reduction
Project risk	There is a strong belief in the ability of structure, planning and control mechanisms to manage and contain risks. Moreover there is a recognition that the incremental approach can result in requirements that are "inconsistent or internally-conflicting" (DSDM 2013a) which is correctly identified as a project rather than a requirements risk
Supplier risk	DSDM offers little in the way of structural constraints for external suppliers to become involved in projects. Although risk management advice is not part of the mainstream literature there is an acknowledgement of the contractual environment in which suppliers exist and their associated risks. Specifically, there is advice concerning the outsourcing of projects (DSDM 2003b)
People risk	This risk is seldom cited though it could be argued that the principles of collaboration and continuous and clear communication alleviate to some extent the risks that might arise here. On the whole, however, this risk does not feature prominently

DSDM artefacts comfortably accommodate risk tailoring, thus enabling us to provide some illustrations of what might be useful in practice (see Table 28). If the project context is embedded in a wider enterprise risk management environment, then it is advisable that the pre-project phase acknowledges this in the project *Terms of Reference* document. The recording of high level risks in the Feasibility and Foundations phases we alluded to earlier ought to be recorded in the risk list and the *Feasibility Assessment* document and used as a quality gate mechanism to decide if the project should be undertaken and if it is aligned with the wider corporate risk appetite. According to DSDM, conclusion of the *Foundations* phase results in the production of the documents listed in Table 28 which can easily be used to contribute to an overall understanding of risk management.

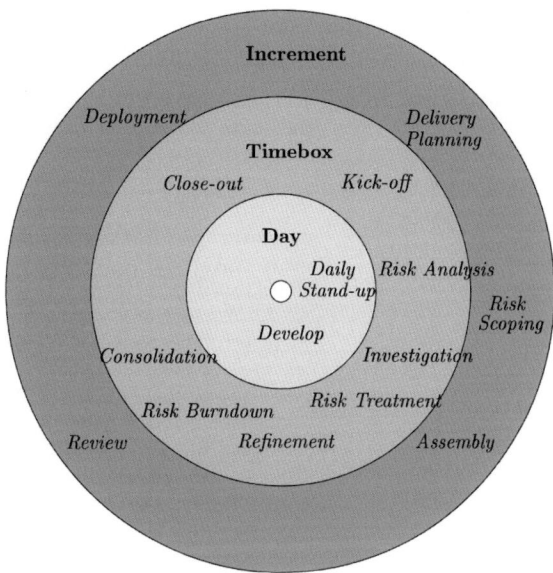

Risk Management

DSDM already accommodates, albeit in a narrow frame, many elements of risk management. Specific artefacts exist that directly address risks (e.g., risk log, delivery plan) and its practices support the process of identifying, analysing and treating risks (e.g., timebox). Some practices are frequently mentioned as a means of tackling one or more risks whilst others provide useful contact points for discussing risks. For example, prototyping is explicitly cited as being appropriate "where significant business or technical uncertainty exists" (DSDM 2012) whereas daily stand-ups offer the opportunity to discuss issues and risks (DSDM 2013b). Table 29 summarizes the response strategies as they appear in the DSDM literature.

The practices described in the Chapter "Agile Risk Management" (e.g., risk stories, scoring and tagging) could easily be deployed in a DSDM environment and accordingly we will not repeat our advice here.

Conclusions

Of all the methodologies in this book DSDM exhibits the greatest commitment to risk management and despite its relatively narrow focus there is a sense that risk consideration permeates its practice. Much of the risk management technique found in DSDM borrows heavily from traditional project risk management and one cannot help feeling that it has simply been grafted on top of the framework with only modest consideration of agile principles. Its acceptance might therefore owe more

Table 28 Risk management comments on the Foundation phase documentation

Document	Risk Management Amendments
Business foundations	Inclusion of the risk appetite and drivers together with a commentary on their alignment with the wider enterprise risk management framework where applicable. High level risks in terms of threats and opportunities to the business should also be listed here
Management foundations	An assessment of the capability of the organization and the maturity of application of DSDM belong to assessment of the project and people risks. This can be extended to the supplier and partner network to address supplier risks (e.g., ability to deliver to required quality standards, off-setting of internal risk through the use of outsourcing etc.). Moreover, deployment and delivery plans can be structured to better manage schedule and supplier risk. Finally once the project is complete an assessment can be made of the project risk management process itself in the *Project Review Report* and the *Benefits Assessment* documents. Note that DSDM already cites the use of this document to record risks (DSDM 2010, 2013a) so our advice merely adds to existing practices
Solution foundations	Identification of which agile techniques are most appropriate for the project and how they can be employed to tackle risks (e.g., applicability, effectiveness, efficiency). For example, facilitated workshops and modelling become more important when the requirements and technical risk are high

Table 29 DSDM interpretation of the classical risk strategies

Risk Strategy	Application
Accept	This strategy is mentioned on occasion and can be reasonably assumed to be part of the risk repertoire
Exploit	This strategy is seldom if ever mentioned as the focus is, by definition, on the handling of negative risks
Share	The consciousness of the contractual environment and the issues that arise in integrating agile practices makes possible the inclusion and management of third parties which is essential to implementing this strategy
Transfer	The same conditions that support the sharing strategy also underpin the transfer strategy though this strategy is seldom if ever explicitly mentioned in the DSDM literature. Contractual matters in relation to outsourcing are, however, addressed in (DSDM 2003b)
Reduce	As to be expected this is the strategy that is most commonly cited in the DSDM literature and several sources cite concrete measures and how best to apply them (DSDM 2010, 2013a, b)
Avoid	This strategy is rarely explicitly mentioned but there are hints that it is subsumed under the reduction strategy

to it being a familiar sight to the target audience of the methodology rather than it being inherently a reinterpretation or adaptation of risk management practices to an agile world.

Concluding Remarks

There is ample scope for using agile risk management in the core methodologies outlined in this book. Moreover, the practices described in the Chapter "Agile Risk Management" (in particular risk walling) ought to add value without being intrusive. Whilst each methodology has its own understanding of and contribution to the tackling of risks, it is clear that DSDM exhibits the most comprehensive approach to risk management owing in large part to its holistic attitude towards the development and delivery of solutions. Adapting and extending any methodology even with the simplest of measures is, however, likely to bring immediate benefits to risk awareness, communication and management.

Enterprise Agility

Abstract We take a glimpse at the growing interest in agility applied to the enterprise where projects assume greater dimensions in terms of team sizes, geodispersion and complexity. We reflect on the contribution in this respect of the frameworks already discussed in this book (e.g., DSDM) together with some new arrivals (e.g., DAD, SAFe). Although we acknowledge the coming of age of agility, we note the absence of reference to, or integration of, enterprise risk management though we are optimistic that this will be rectified in the coming years.

Overview

Over time agility has evolved from a discipline that has focused entirely on engineering and product development into one that seeks to embrace the entire solution delivery process from initiation through to construction and deployment of working software replete with changes to underlying business process and organizational structures. We use the term *horizontal scaling* to describe the widening of the scope of agility beyond the traditional focus on solicitation of requirements, construction of solutions and building of deployment artefacts and the term *vertical scaling* to denote to the same agile practices but undertaken on a larger scale (e.g., use of multiple teams in concert, increased complexity, wide geographic dispersion). Since the early literature which first considered how agility might scale (Eckstein 2004) there has emerged several contenders for the title of enterprise agile framework, including Disciplined Agile Delivery (DAD) and the Scaled Agile Framework (SAFe) which we introduce in this chapter. Most of these synthesize practices from a wide range of other methodologies.

Enterprise agility endeavours to go beyond the product development and project management boundaries that grew out of teams focused on delivery of specific stakeholder value. The scope of such agile practices has typically been delimited by the expression of a business case (e.g., vision, requirements) to the delivery of a working solution (e.g., software artefacts, user and operations documentation) but

A. Moran, *Agile Risk Management*, SpringerBriefs in Computer Science, DOI: 10.1007/978-3-319-05008-9_5, © The Author(s) 2014

Fig. 36 Methodologies in terms of scope and scalability. Published with kind permission of © Alan Moran 2013. All Rights Reserved

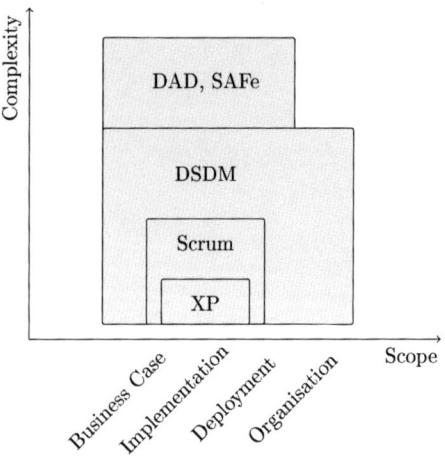

seldom extended to matters of project governance (i.e., ensuring effective delivery of a solution) or adaptation of organisational structures or processes. These latter details were historically addressed by relatively few methodologies (e.g., DSDM). That delivery of a solution usually involves so much more than just the creation of an artefact is clearly evident from the plethora of activities such as infrastructure upgrades, data migration, training, early life support, changes to working practices and decommissioning. Sometimes the organisation itself is the impediment in this respect by segmenting functions in such a manner that certain activities cannot be performed by the project team. An example of this is the COBIT PO4.11 "Segregation of Duties" that is justified on efficiency grounds and is claimed to reduce risk in relation to damage and compliance (ISACA 2012a). Moreover the assumption that teams are small, co-located and highly communicative (e.g., access to customer representatives) with less focus on geographically distributed large teams (e.g., hundred of developers spread over several countries) or those that work on highly complex projects (i.e., in terms of domain, technology, organisation or regulations) was prevalent. Over time, however, research and literature has emerged concerning the scalability of agility (Bannerman et al. 2012; Sutherland et al. 2007; DSDM 2003b; Lines and Ambler 2012; Leffingwell 2007) as well as the growing role of risk management therein (Hossain 2009; DSDM 2003c) demonstrating that the era of enterprise agility is upon us.

From Fig. 36 we see the scope of project activities as being initiated by a business case which upon implementation is delivered into the organisation on which it may have some degree of impact (e.g., changes to structures or processes). This in turn may require some oversight in the form of project governance which may stipulate that certain practices must be undertaken (e.g., quality, risk or configuration management). On the other hand agile projects may range from small co-located teams to large distributed team structures that require product segmentation into independent components and appropriate coordinating structures. Furthermore complexity may vary significantly spanning organisational, technical, domain or regulatory boundaries.

Within this context we place XP and Scrum firmly in the product development scope and though we acknowledge that a certain amount of vertical scalability is achievable with Scrum of Scrums structures (see below) we observe little evidence of Scrum extending beyond product development. DSDM, on the other hand, bears testimony to a methodology that is capable of operating on a wide scale of complexity and size and embraces many of the elements of governance that would be expected of enterprise agility. Finally DAD and SAFe (both of which are described later in this chapter) are specifically aimed at the enterprise and borrow heavily from other methodologies to cover product development and project management concerns. It is here, as can also be observed with DSDM, that we also gain some insight into what it might mean for agility to be said to be "enterprise aware".

In risk management terms there is a shift as a project deliverable transitions into production away from anticipatory and reactive measures towards a controls based approach wherein tolerances are defined and thresholds determine the point at which action must be undertaken (e.g., key risk indicators). This reflects the transfer of focus away from production concerns towards operational matters where the project artefact is perhaps one of many which together share the same risk environment and must collectively contribute towards smooth operations. There already exists extensive literature describing how risk is managed in this context (ISACA 2009a, b) including IT service management (OGC 2011a) and IT governance (ISACA 2012a) frameworks as well as guidance on how agile methodologies can be integrated into such environments (DSDM Consortium 2010a). Emergent risk issues (e.g., security, compliance) also arise in this context since they are typically difficult to treat adequately at the product or sometimes even at the project level (e.g., if the project is one element in a larger portfolio).

Scrum of Scrums

Scrum, understood in the strictest sense, imposes a number of structural constraints (e.g., the number of team members, time allocation for daily stand-ups) that inhibit scalability. It is suggested that a degree of vertical scaling can be achieved by creating a number of teams each of which adhere to traditional Scrum practices, all of whom send designated individuals to convene in order to coordinate intra-team affairs in their own daily stand-up (coined a "Scrum of Scrums"). Canonical advice on how such meetings should be conducted is available (Cohn 2007).

Scrum advocates also argue the case for geographical vertical scaling (Sutherland et al. 2007) which makes significant claims about the productivity of a geographically distributed team of over fifty developers.[1] It has even been suggested that the Scrum of Scrums approach can be extended to agile programme management (Zarnett 2012)

---

[1] Recall that Scrum uses the term "developer" to denote any member of a team that is neither a Scrum Master nor a Product Owner irrespective of the actual duties they perform.

where it is envisaged as a supporting aid. These views are confirmed by independent research which notes that agility in general and Scrum in particular continue to deliver competitive advantage in spite of the violation the co-location principle implicit in its model (Bannerman 2012) though an earlier study hints at possible limitations whilst admitting that the jury might still be out on the Scrum verdict (Hossain 2009).

One example of how Scrum was deployed on a large scale in an enterprise setting is that of saleforce.com which in 2006 undertook a radical "big bang" roll-out of Scrum across the organization in a bid to halt the increasing bureaucracy and to return to its core values (Emerald 2012). The programme was completed in only three months achieving the following impressive results:

> During the first year of making the switch, Salesforce.com released 94 % more features, delivered 38 % more features per developer and delivered over 500 % more value to their customers compared to the previous year. (Cohn 2009)

Central to this success was not only executive support but also a strong empowered change team, an adherence to principles rather than the mechanics of how they are implemented and a commitment to complete transparency, which has been cited as a risk mitigation factor (Denning 2011) during the agile transformation. There was also an explicit focus on "being agile" rather than "doing agility" heralding a new way of thinking for the company rather than merely introducing a new business process. Finally a huge investment in training and coaching rounded off the package.

In spite of the apparent scalability success of Scrum it must be noted that since the core traits of the model are restrained, the scope for horizontal scaling is rather limited as might be the number of vertical scaling factors with which it can cope. In the case of salesforce.com, for example, an important lesson was the adaptation of the methodology based on experiences already gathered by the company (Denning 2011). We observed already that other frameworks express a willingness to subsume Scrum into their models and enrich it with features that are absent from Scrum (DSDM Consortium 2012b) and as we shall see later this eagerness can also be found in the more recent entrants (Lines and Ambler 2012; Leffingwell 2007).

Dynamic Systems Development Method

Owing to its embracing of the entire solution development process, Dynamic Systems Development Method (DSDM) already exhibits many of the characteristics expected of horizontally scaled frameworks (Stapleton 2003; DSDM 2010). Of those methodologies that were already established at the time of writing of the agile manifesto, it was probably the most complete in terms of the full scope of the delivery process extending beyond the mere production method as illustrated by artefacts such as evolving and deployed solutions. Moreover its advocates testify to its vertical scalability citing case studies involving complexity and team sizes (DSDM 2003a). Since the emergence of AgilePM® it has become recognised as a leading agile project methodology that displays a panoramic enterprise awareness (e.g., attention to quality, configuration and risk management). Beyond the method itself, however, there is a treasure trove of whitepapers offering advice on a broad range of enterprise integration issues (DSDM 2004, 2003c, b, 2006) and a whole body of knowledge concerning the application of DSDM to portfolio management and to large projects (DSDM Consortium 2012a; DSDM 2003a). Thus it is fair to say that not only DSDM itself but also the wider context in which it is embedded demonstrate considerable enterprise maturity.

In some cases DSDM can be seen to be applied in concert with other methodologies and is organisational structural focus could be said to complement features of other approaches (e.g., the architectural emphasis found in SAFe). For example, one case study citing the UK Highways Agency describes the integration of PRINCE2® and DSDM where the former covers project governance and management perspectives whilst the latter applies agile software product development techniques to deliver the solution (DSDM Consortium 2010b).

Disciplined Agile Delivery

Disciplined Agile Delivery is a process framework that draws on the core methodologies discussed in this book (i.e., XP, Scrum and DSDM) together with elements of other methodologies (notably Agile Modelling, Agile Data, Kanban and Unified Process) and endeavours to create out of these a platform for horizontal and vertical scalability. Indeed it is very explicit about the nature of this scaling citing team size, geographic distribution, regulatory compliance environments, domain, technical and organizational complexity amongst its scaling factors. Playing on subtle nuances (e.g., solution delivery rather than working software) it attempts to stretch the natural boundaries of agility describing itself in the following terms

> The Disciplined Agile Delivery (DAD) process framework is a people-first, learning-oriented hybrid agile approach to IT solution delivery. It has a risk-value lifecycle, is goal-driven, is scalable and is enterprise aware (Lines and Ambler 2012).

Its understanding of risk focuses understandably on business (which is effectively a combination of requirements and schedule risk as described in this work) and technical aspects but also includes operational, process and organizational risk though it frames these exclusively in negative terms as "an exposure to a potentially negative outcome" (Lines and Ambler 2012). Reference to a "risk-value lifecycle" is reminiscent of the work of Boehm (Boehm and Turner 2009) though its treatment is less detailed. Finally whilst enterprise awareness is commendable and will doubtless endear the methodology to its target audience, from a risk management perspective there is little connection with notions of enterprise risk management already commonplace in large organizations (Moeller 2011; ISACA 2009a) though one might reasonably assume that traces thereof are to be found in practice.

Streamlining ideas derived from Unified Process, DAD is divided into inception (e.g., funding, vision, scoping, planning), construction (e.g., balancing risk and value, modelling, implementation, planning) and transition (e.g., production readiness

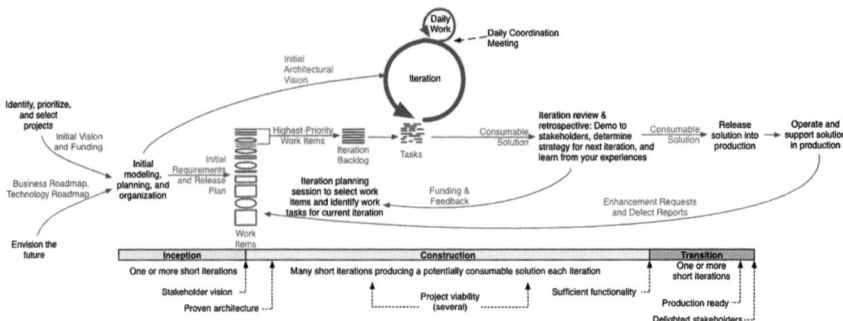

Fig. 37 Basic DAD model based on scrum. © Scott Amber + Associates 2013. Reprinted with Permission

assessment, deployment) phases that easily accommodates and extends the Scrum model (as illustrated in Fig. 37) as well as other forms of solution development (e.g., Lean). Indeed its attention to both agile and lean communities affords it more scope than one might ordinarily associate with a framework of this nature.

DAD teams illustrate the practical nature of scalability. For example small teams typically comprise of a technical lead (which can assume architectural responsibility), product owner and other members supported by technical and domain experts outside of the project team. Medium sized teams, which focus on feature or component development, maintain separate roles and augment them with specialists whilst retaining the "supporting cast" of experts. Finally large teams mirror Scrum of Scrums practices by aggregating specific roles across the individual teams into team groups (e.g., project management team, product owner team, architecture owner team).

DAD also embodies a decision making framework, the Software Development Context Framework (SDCF), that defines the selection and tailoring of software development strategies that reflect the specifics of the project circumstances (Ambler 2013c). It uses selection factors such as team skills, culture, nature of problem and related business constraints together with scaling factors based on size, geographic distribution and complexity to arrive at a team organisation, delivery process and tooling appropriate for the project. Risk consideration is embedded in the SDCF in so far as people and supplier risks are accounted for and the risk-value lifecycle is applied throughout. DAD continues to be developed borrowing on notions taken from Enterprise Unified Process (Ambler 2013b; Ambler et al. 2005), which takes a lifecycle view of system solution development, and will likely embrace enterprise architecture and portfolio management reflecting wider trends in the agile community. Accordingly it is to be expected that risk management may feature more prominently in the future, perhaps assuming the features of governance together with balance of risk and reward currently associated with enterprise risk management.

Scaled Agile Framework

The Scaled Agile Framework (SAFe) describes itself as an "interactive knowledge base for implementing agile practices at enterprise scale" (SAFe 2013) and cites a number of scalable practices that must be mastered in order to achieve enterprise agility. These include the definition and organization of committed teams tasked with building and testing components that accountable for delivering results with dual levels of planning and tracking (akin to our separation of iterative development and incremental deployment), a mastery of the iteration as the "heartbeat of agility" (Leffingwell 2007), smaller and more frequent releases (in tune with lean thinking), concurrent testing and integration and a culture of reflection and adaptation.

There is an emphasis on the atomic nature of requirements and design that makes a strong argument for their continual evolution during the development process on grounds that "neither has any meaning without the other" (Leffingwell 2007) rather than separating them as distinct phases. Institutionalization of continuous integration is supported by efficient project structures that draw heavily on XP practices. This reflects a wider appreciation within the agile community of continuous feedback that today has come to be understood as involving much more than just testing (e.g., automated code analysis and security reviews).

Proposing a hybrid matrix organisation, the case is made for teams that draw the right people from disparate functions and departments (e.g., product definition, software development, testing and quality) to achieve an agile organisation of staff. This is reflected in the "servant leadership" role of team leads who act as facilitators empowering those within their teams. Organisational structures range from Scrum of Scrums who track progress on a daily basis to heterogeneous steering committees comprising not only of team leads but also architects, managers and other stakeholders as appropriate (Fig. 38).

SAFe challenges the notion that agility is confined to the realm of small co-located teams based purely on emergent architecture that are light on requirements analysis. Rather it suggests that at scale, "all development is distributed development" (Leffingwell 2007) and that agility must therefore rise to this challenge by scaling it practices and incorporating specific techniques (e.g., forward-looking architecture referred to as an "architecture runway"). It acknowledges the impediments that some organisations place in the way of agility such as resistance to change born of a desire to protect established assets, introduction of controls that subtly introduce waterfall style gateways and reward systems that favour the individual over the collective.

The practices of SAFe find integration in the creation of an agile enterprise founded on intentional architecture arising from largely independent components with clearly defined interfaces, a lean approach to requirements based on product visions and just-in-time elaborations, employment of functionality variability to coordinate product releases, appropriate tooling to support highly distributed teams, organisational and performance measure adaptations (e.g., agile balanced score cards), some of which have already been alluded to above.

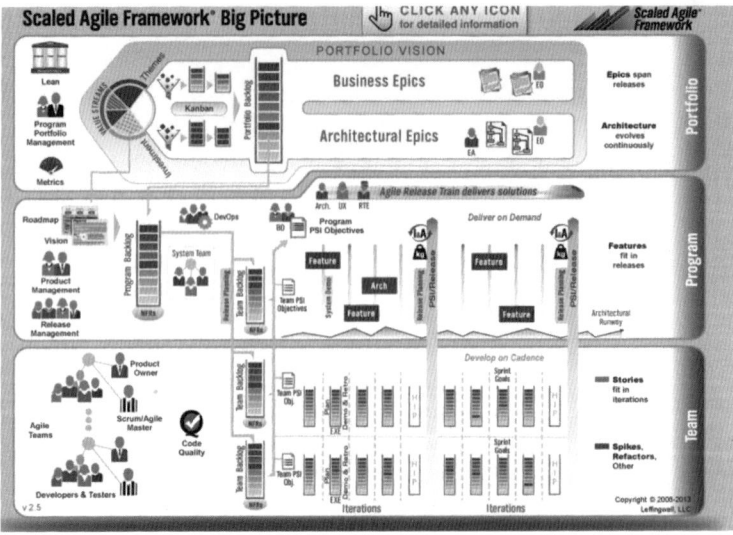

Fig. 38 The SAFe big picture. © Leffingwell LLC 2013. Reprinted with Permission

In terms of risk, SAFe borrows not only the agile risk practices but also the belief that their implicit application suffices to mitigate risk. It may thus lack the necessary framework with which risk and reward can be effectively balanced. Risk, as we have come to expect in agile circles, is typically limited to technical and requirements risk (e.g., the "Yes, But" syndrome described in (Leffingwell 2007)) and an indication of the reactive nature of SAFe can be found in the explanation of why teams might miss their goals owing to unanticipated technical obstacles suggesting that early iterations "will immediately expose the risks, wherever they may lay" (Leffingwell 2007). We suggest that risk scoping might alleviate such issues and that practice of our flow principle may enable projects to continue unabated.

Concluding Remarks

The rise of enterprise agility in recent years reflects the increasing attention agility is receiving. We observe the broad nature of responses within the agile community ranging from integrationist thinking of DSDM to the extension of Scrum based practices and realignment of organisations along agile lines found in DAD and SAFe. Sadly, enterprise risk management, also a strong growth phenomenon in recent years, receives little attention from any of these approaches. Thus it remains unclear how scaled agility maps to the risk components or management objectives described by mainstream ERM frameworks (Moeller 2011; ISACA 2009a). We feel therefore that there is still some way to go before enterprise agility can be said to have incorporated enterprise risk management.

Appendix A
Agile Techniques

The following list of agile techniques reflects those practices (and their interpretation) most commonly encountered by the methodologies in this book. The list should not be understood as exhaustive and we refer the interested reader to the primary sources for more detailed descriptions and advice.

Technique	Description
Heterogeneous small teams	Formation of teams, with preferably five to nine (Miller 1956) co-located members comprising of business representatives and solution developers, with a preference towards generalists over specialists
Product vision	Formulation of a simple evocative statement that clearly states the purpose of the product which the project is endeavouring to create
User stories	Formulation in simple narrative form of requirements together with statements of what constitutes successful implementation (definitions of done)
Coding standards	Adherence to a common standard (perhaps verified by static and runtime code analysis) throughout the codebase
Iterative development	Traversal of the entire SDLC within a fixed timeframe with the aim of producing tested and stable code that contributes towards the evolving solution
Incremental delivery	Release into production environments of partial evolving solutions throughout the project lifecycle
Continuous integration	Continuous merging (and typically unit testing) of code contributions into a shared repository in order to reduce integration hurdles and provide feedback at the code commit level
Refactoring	The restructuring of code internals in a manner that does not alter its outward behaviour.
Pair programming	Working together of two solution developers on the same piece of code at the same terminal often interchanging the roles of implementer and reviewer

(continued)

A. Moran, *Agile Risk Management*, SpringerBriefs in Computer Science, DOI: 10.1007/978-3-319-05008-9, © The Author(s) 2014

(continued)

Technique	Description
Test driven development	Creation of unit tests prior to implementing solutions as part of a culture of continual validation and verification
Prototyping	Creation of a (possibly throwaway) mock to triail the design of a solution or to better understand the problem
Modelling	Creation of a conceptual design (e.g., process diagram) for the purpose of discussion and gaining of consensus concerning a problem or its solution
Facilitated workshop	Structured workshop wherein team members work towards specific goals with the support of an independent facilitator
Daily standup	Brief and focused daily meeting of all team members focusing on what has been achieved since the last meeting, what is planned that day and what is blocking work
Increment planning	Creation of high level (and often imprecise) plans based on available knowledge
Iteration planning	Creation of detailed short term plans (typically based on user stories) complete with estimates and priorities and derived from the increment plan
Timeboxing	Structured interval of time to control to segment and control activities
Planning poker	A variant of the Wideband Delphi method in which consensus is sought during prioritization by requiring participants to vote on and discuss their priorities
MoSCoW prioritization	Classification of priorities in terms of MUST, SHOULD, COULD and WONT (in respect of the current iteration!) and the accommodation of contingency in planning
Kanban	Pull based taskboard (with status swimlanes) where team members assign tasks to themselves and use states to update progress (e.g., "In Progress", "Completed")
Information radiator	A dashboard concept presenting project relevant information at a location to which project members naturally feel drawn (e.g., coffee machine)
Burndown charting	Regularly updated display of project progress (usually in terms of decreasing remaining effort)
Retrospective	Gathering of all team members to discuss lesson learned during the last iteration with a view to identify and act on possible process improvements
Demonstration	Presentation of the state of an evolving solution at the end of an iteration in order to solicit feedback from stakeholders
Sustainable pace	Maintenance of normal working hours and the avoidance of fluctuations and peaks (e.g., as project deadlines approach)

References

Adams, J. (1995). *Risk*. London: UCL Press.

Agile Alliance. (2013). *Guide to Agile practices* (Timeline). Retrieved October 1, 2013, from http://guide.agilealliance.org/timeline.html

Altlassian. (2013). *Agile project management software: Atlassian JIRA agile*. Retrieved October 1, 2013 from https://www.atlassian.com/software/jira-agile/overview

Ambler, S. (2012). *The Agile Unified Process (AUP)*. Retrieved October 1, 2013, from http://www.ambysoft.com/unifiedprocess/agileUP.html

Ambler, S. (2013a). *Agile adoption strategies: November 2011 survey results*. Retrieved December 1, 2013, from http://www.ambysoft.com/surveys/agileStateOfArt201111.html

Ambler, S. (2013b). *Enterprise unified process (EUP): Strategies for enterprise Agile*. Retrieved December 1, 2013, from http://enterpriseunifiedprocess.com/

Ambler, S. (2013c). *Scaling Agile: The software development context framework*. Retrieved December 1, 2013, from http://scaledagileframework.com/

Ambler, S. (2013d). *Surveys exploring the current state of information technology practices*. Retrieved December 1, 2013, from http://www.ambysoft.com/surveys/

Ambler, S. (2003). *Agile database techniques: Effective strategies for the Agile software developer*. Ambler: Wiley.

Ambler, S., Nalbone, J., & Vidoz, M. (2005). *The enterprise unified process: Extending the rational unified Process*. Englewood Cliffs: Prentice Hall.

Anderson, D. (2003). *Agile management for software engineering: Applying the theory of constraints for business results*. Englewood Cliffs: Prentice Hall.

Arnuphaptrairong, T. (2011). Top ten lists of software project risks: Evidence from the literature survey. *Proceedings of the International MultiConference of Engineers and Computer Scientists, 1*, 16–19.

AS/NZS. (2004). *Australian and New Zealand Risk Management Standard AS/NZS 4360:2004*, Standards Australia.

Augustine, S. (2008). *Managing Agile projects*. Harlow: Prentice Hall.

Bannerman, P. et al. (2012). Scrum practice mitigation of global software development coordination challenges: A distinctive advantage? *45th Hawaii International Conference on System, Science* (pp. 5309–5318).

Beck K., et al. (2001a). *Manifesto for Agile software development*. Retrieved October 1, 2013, from http://agilemanifesto.org/

Beck K., et al. (2001b). *Principles behind the Agile manifesto*. Retrieved October 1, 2013, from http://agilemanifesto.org/principles.htm

Beck, K. (2002). *Test driven development by example*. Boston: Addison-Wesley.

Beck, K. & Andres, C. (2004). *Extreme programming explained: Embrace change*. Boston: Addison-Wesley.

A. Moran, *Agile Risk Management*, SpringerBriefs in Computer Science,
DOI: 10.1007/978-3-319-05008-9, © The Author(s) 2014

Benington, H. (1983). Production of large computer programs. *IEEE Annals of the History of Computing (IEEE Educational Activities Department)*, 5, 350–361.

Boehm, B. (1986). A spiral model of software development and enhancement. *ACM SIGSOFT Software Engineering Notes*, *11*, 14–24.

Boehm, B. (1988). A spiral model of software development and enhancement. *IEEE Computer*, *21*, 61–72.

Boehm, B. (1991). Software risk management: Principles and practices. *IEEE Software*, *8*, 32–41.

Boehm, B. & Turner, R. (2003). Using risk to balance agile and plan-driven methods. *IEEE Computer Society*, *36*, 57–66.

Boehm, B. & Turner, R. (2009). *Balancing agility and discipline: A guide for the perplexed*. Boston: Addison-Wesley.

Cockburn, A. (2007). *Agile software development: The cooperative game*. Boston: Addison-Wesley.

Cohn, M. (2007). *Advice on conducting the scrum of scrums meeting*. Retrieved October 1, 2013, from http://www.scrumalliance.org/community/articles/2007/may/advice-on-conducting-the-scrum-of-scrums-meeting

Cohn, M. (2010). *Agile estimation and planning*. New Jersey: Prentice Hall.

Cohn, M. (2009). *Succeeding with Agile: Software developing using scrum*. Boston: Addison-Wesley.

Collier, P. (2009). *Fundamentals of risk management for accountants and managers: Tools and techniques*. London: Elsevier.

Connelly, K. et al. (2010). *The growing role of the board in risk oversight*. Retrieved October 1, 2013, from https://www.spencerstuart.com/research-and-insight/the-growing-role-of-the-board-in-risk-oversight

Coplien, J. & Harrison, N. (2005). *Organizational patterns of Agile software development*. Upper Saddle River: Pearnson Prentice Hall.

COSO (2004). *COSO Enterprise Risk Management-Integrated Framework*. Committee of Sponsoring Organizations of the Threadway Commission.

Craddock, A. (2013). *Agile Project Management and Scrum (Pocketbook)*. DSDM Consortium.

Denning, S. (2011). Successfully implementing radical management at Salesforce.com. *Strategy and Leadership*, *39*, 4–10.

Derby, E. & Larsen, D. (2009). *Agile retrospectives: Making good teams great*. Upper Saddle River: Pragmatic Bookshelf.

DSDM. (2003a). *The application of DSDM to large projects*. DSDM Consortium.

DSDM. (2003b). *Whitepaper outsourcing DSDM projects*. DSDM Consortium.

DSDM. (2003c). *Whitepaper risk management*. DSDM Consortium.

DSDM. (2004). *Whitepaper organisation suitability filter*, DSDM Consortium.

DSDM. (2006). *Whitepaper DSDM and changing business processes bringing people, process and technology together*. DSDM Consortium.

DSDM. (2010). *Agile project management handbook*. DSDM Consortium.

DSDM. (2012). *The DSDM agile project framework (Pocketbook)*. DSDM Consortium.

DSDM. (2013a). *AgileBA: The handbook for business analysts*. DSDM Consortium.

DSDM. (2013b). *The Agile project management framework: Timeboxing*. DSDM Consortium.

DSDM Consortium. (2010a). *Agile project and service management: delivering IT services using ITIL, PRINCE2 and DSDM atern*. London: Stationery Office Books.

DSDM Consortium. (2010b). *An agile approach to software systems development for the highways agency*. Retrieved October 1, 2013, from http://dsdm.org/sites/default/files/An-Agile-Approach-to-Software-Systems-Development-for-the-Highways-Agency-Sept-10-V3.pdf

DSDM Consortium. (2012a). *The Agile PMO (Pocketbook)*. DSDM Consortium.

DSDM Consortium. (2012b). *The DSDM Agile project framework for scrum*. http://live-dsdm.gotpantheon.com/dig-deeper/white-papers/dsdm-agile-project-framework-scrum

Eckstein, J. (2004). *Agile software development in the large: Diving into the deep*. Boston: Pearson Education.

Emerald, J. (2012). Salesforce.com beats the entropic effect. *Strategic direction*, *28*, 26–28.

Garland, D. & Fairbanks, G. (2010). *Just enough architecture: A risk-driven approach*. Boulder: Marshall & Brainerd.

Gilb, T. (2005). *Competitive engineering: A handbook for systems engineering, requirements engineering, and software engineering using planguage*. London: Butterworth-Heinemann.

Hazzan, O. & Dublinsky, Y. (2008). *Agile software engineering*. London: Springer Verlag.

Hikaka, M. et al. (2005). *Is extreme programming just old wine in new bottles: A comparison of two cases*. Retrieved October 1, 2013 from http://www.sirel.fi/ttt/Downloads/Old%20Wine%20in%20A%20New%20Bottle.pdf

Hillson D. (2009). *Managing risk in projects*. Gower.

Hofstede, G. (2013). *Dimensions-Geert hofstede*. Retrieved October 1, 2013, from http://geert-hofstede.com/dimensions.html

Hofstede, G. (2003). *Culture's consequences: Comparing values, behaviors, institutions and organizations across nations*. Oaks CA: SAGE Publications.

Holling, C. (1979). Myths of ecological stability. *Studies in crisis management*.

Horwath, C. (2011). *Risk appetite & tolerance*. London: Institute of Risk Management.

Hossain, E. et al. (2009). Risk identification and mitigation processes for using scrum in global software development: A conceptual framework, *16th Asia-Pacific, Software Engineering Conference* (pp. 457–464).

IBM. (2011). *Disciplined agile delivery: an Introduction*. Retrieved October 1, 2013, from https://www.ibm.com/developerworks/community/blogs/ambler/entry/disciplined_agile_delivery_an_introduction_white_paper22?lang=en

IBM. (2012). *Form 10-k, annual report: International business machines corporation*. http://www.sec.gov/Archives/edgar/data/51143/000104746913001698/a2212340z10-k.htm

IRM et al. (2002). *A risk management standard*. IRM/ALARM/AIRMIC.

ISO. (2002). *ISO/IEC Guide 73 risk management-Vocabulary-Guidelines for use in standards*. International Standards Organization.

ISO. (2009). *ISO 31000:2009 risk management-Principles and guidelines*. International Standards Organization.

ISACA. (2009a). *The risk IT framework*. ISACA.

ISACA. (2009b). *The risk IT practitioner guide*. ISACA.

ISACA. (2012a). *COBIT 5: A business framework for the governance and management of enterprise IT*. ISACA.

ISACA. (2012b). *COBIT 5 for information security*. ISACA.

Jaana Nyfjord, J. & Kajko-Mattsson, M. (2007). Commonalities in risk management and agile process models. *In Proceedings of the 2nd International Conference on Software Engineering Advances*.

Jaana Nyfjord, J. & Kajko-Mattsson, M. (2008). Outlining a model integrating risk management and Agile software development. *Software Engineering and Advanced Applications*, 476–483.

Jeffries, R. (2001). *What is extreme programming? Whole team*. Retrieved October 1, 2013, from http://xprogramming.com/what-is-extreme-programming/#whole

Jeffries, R. (2006). *Risk management*. Retrieved October 1, 2013, from http://xprogramming.com/articles/risklog/

Jeffries, R. (2013). *XProgramming*. Retrieved October 1, 2013, from http://xprogramming.com/index.php

Knight, F. (1921). *Risk, Uncertainty and Profit*. Boston: Houghton Mifflin.

Krebs, J. (2008). *Agile portfolio management*. WA, USA: Microsoft Press.

Larman, C. (2003). *Agile & Iterative development: A manager's guide*. Boston: Addison-Wesley.

Leffingwell, D. (2007). *Scaling Software Agility: Best Practices for Large Enterprises*.

Li, M. et al. (2006). *A Risk-Driven Method for eXtreme Programming Release, Planning*, pp. 423–430.

Lines, M. & Ambler, S. (2012). *Disciplined Agile dlivery: A practitioner's guide to Agile software delivery in the enterprise*, IBM Press.

Loukides, M. (2012). *What is DevOps?*. O' Reilly & Associates.

Martin, J. (1991). *Rapid application development*. USA: Macmillan.

McBreen, P. (2002). *Questioning eXtreme programming*. New York: Addison-Wesley.

McKeown, C. & Holmes, A. (2009). *English dictionary*. Glasgow: HarperCollins Publishers Limited.

Miller, G. (1956). The magical number seven, plus or minus two: Some limits on our capacity for processing information. *Psychological Review, 63*(2), 81–97.

Moeller, R. (2011). *COSO Enterprise risk management: Establishing effective governance, risk and compliance processes*. New York: Wiley.

Murray-Webster, R. & Hillson, D. (2008). *Managing Group Risk Attitude*. Gower.

Nelson, C., Taran, G.l., & de Lascurain Hinojosa, L. (2008). Explicit risk management in Agile processes (pp. 190–201). *Agile processes in software engineering and extreme programming*. Berlin: Springer.

Nyfjord, J. (2008). *Towards integrating agile development and risk management*. Stockholm: Ph.D. thesis.

Office of Government Commerce (2007). *Management of Risk: Guide for Practitioners*. London: The Stationery Office.

OGC (2009). *Managing successful projects with PRINCE2*. London: Stationery Office Books.

OGC (2011a). *ITIL V3 Foundation Handbook*. London: Stationery Office Books.

OGC (2011b). *Managing successful programmes*. London: Stationery Office Books.

Palmer, S. & Felsing, J. (2002). *A practical guide to feature-driven development*. Upper Saddle River: Prentice Hall.

Patton, J. (2008). *Don't know what I want, but I know how to get it*. Retrieved December 1, 2013, from http://www.agileproductdesign.com/blog/dont_know_what_i_want.html

Pichler, R. (2010). *Agile product management: Creating products that customers love*. Boston: Addison-Wesley.

Poppendieck, M. & Poppendieck, T. (2003). *Lean software development: An agile tookit*. Boston: Addison-Wesley.

Project Management Institute (2009). *Practice standard for project risk management*. Project Management Institute, Incorporated.

Project Management Institute. (2013). *A guide to the project management body of knowledge: PMBOK guide*. Incorporated: Project Management Institute.

Protiviti, (2013). *Guide to enterprise risk management*. Retrived October 1, 2013, from http://www.protiviti.com/en-US/Pages/Guide-to-Enterprise-Risk-Management.aspx

Ropponen, J. & Lyytinen, K. (2000). Compoents of software development risk. *IEEE Transactions on Software Engineering, 26*.

Royce, W. (1970). Managing the development of large software systems. *Proceedings of IEEE WESCON, 26*.

Saaty, T. (1994). *Fundamentals of decision making and priority theory with the analytic hierarchy process*. Singapore: Rws Pubns.

SAFe. (2013). *Scalable agile framework*. Retrieved October 1, 2013, from http://scaledagileframework.com/

Schwaber, K. (2004). *Agile project management with scrum*. Rochelle Park: Microsoft Press.

Schwaber, K. (2006). *Scrum is hard and disruptive*. Retrieved October 1, 2013, from http://www.controlchaos.com/storage/scrum-articles/Scrum%20Is%20Hard%20and%20Disruptive.pdf

Schwaber, K. & Beedle, M. (2001). *Agile software development with scrum*. Boston: Prentice Hall.

Schwarz, M. & Thompson, M. (1990). *Divided we stand: redefining politics, technology and social choice*. Upper Saddle River: Harvester Wheatsheaf.

Sitkin, S. & Pablo, A. (1992). Reconceptualizing the determinants of risk behavior. *The Academy of Management Review, 17*, 9–38.

Stapleton, J. (2003). *DSDM: business focused development*. Upper Saddle River: Pearson Education.

Stephen J., et al. (2011). *System error: Fixing the flaws in government IT*. http://www.instituteforgovernment.org.uk/sites/default/files/publications/System%20Error.pdf

Sutherland, J. (2010). *Iterative vs. Incremental development.* Retrieved October 1, 2013, from http://scrum.jeffsutherland.com/2010/01/iterative-vs-incremental-development.html

Sutherland, J. & Schwaber, K. (2013). *The scrum guide. The definitive guide to scrum: The rules of the game.* Retrieved October 1, 2013, from https://www.scrum.org/Scrum-Guides

Sutherland, J. et al. (2007). Distributed Scrum: Agile Project Management with Outsourced Development Teams. *Proceedings of the 40th Hawaii international conference on system sciences.*

Takeuchi, H. & Nonaka, I. (2013). *The knowledge creating company.* Oxford: Oxford University Press.

Takeuchi, H. & Nonaka, I. (1986). New new product development game. *Harvard Business Review, 96116,* 137–146.

Thompson, M. et al. (1990). *Cultural theory.* Boulder: Westview.

Treasury, H. M. (2004). *The orange Book: Management of risk-principles and concepts.* London: Stationery Office Books.

Versionone (2012). 7th Annual State of Agile Development Survey.

Wells, D. (1999a). *Fix XP when it breaks.* Retrieved October 1, 2013, from http://www.extremeprogramming.org/rules/fixit.html

Wells, D. (1999b). *The rules of extreme programming.* Retrieved October 1, 2013, from http://www.extremeprogramming.org/rules.html

Wells, D. (1999c). *When should extreme programming be used?* Retrieved October 1, 2013, from http://www.extremeprogramming.org/when.html

Wells, D. (2000). *Extreme programming project.* Retrieved October 1, 2013, from http://www.extremeprogramming.org/map/project.html

Wells, D. (2009a). *Extreme programming: A gentle Introduction.* Retrieved October 1, 2013, from http://www.extremeprogramming.org

Wells, D. (2009b). *Iterative planning.* Retrieved October 1, 2013, from http://www.agile-process.org/iterative.html

Williams, L., et al. (2000). Strengthening the case for pair programming. *Software IEEE, 17,* 19–25.

Zarnett, B. (2012). *Running the scrum-of-scrums: Agile program management.* http://www.scrumalliance.org/community/articles/2012/march/running-the-scrum-of-scrums-agile-program-manageme

Printed in Great Britain
by Amazon